Memoirs of the American Mathematical Society
Number 180

Joseph A. Wolf

Unitary representations
of maximal parabolic subgroups
of the classical groups

Published by the
AMERICAN MATHEMATICAL SOCIETY
Providence, Rhode Island

VOLUME 8 · NUMBER 180 (end of volume) · NOVEMBER 1976

Abstract

    All the irreducible unitary representations are found, in an explicit way, for the maximal parabolic subgroups in the various classical series of real and complex Lie groups. In each case, the nilradical is similar to the Heisenberg group, and its representations come out of the Kirillov orbit method. Then the representations of the parabolic subgroup are worked out from Mackey's little group method. The little group usually belongs to a different classical series -- but with smaller matrices -- so the end result in each series is a recursive statement involving several series.

AMS(MOS) subject classification numbers (1970): Primary 22E45.
Secondary 20G05, 20G20, 20G45, 22E10, 22E15, 22E70.

Key words and phrases: Unitary representation, classical group, parabolic subgroup.

Library of Congress Cataloging in Publication Data   **CIP**

Wolf, Joseph Albert, 1936-
  Unitary representations of maximal parabolic subgroups of the classical groups.

  (Memoirs of the American Mathematical Society ;
no. 180)
  Bibliography: p.
  1. Lie groups. 2. Representations of groups.
3. Linear algebraic groups. I. Title. II. Series:
American Mathematical Society. Memoirs ; no. 180.
QA3.A57 no. 180 [QA387] 510'.8s [512'.2] 76-44397
ISBN 0-8218-2180-6

Contents

iii

§0. Introduction.

In this paper we write out the irreducible unitary representations of
the maximal parabolic subgroups of the real and complex classical Lie groups.

Parabolic subgroups appear in automorphic function theory and riemannian
symmetric space theory as normalizers of boundary components, in harmonic
function theory and the theory of homogeneous kaehler manifolds as stability
groups at a point, in quantum field theory and cosmology as the Poincaré
group with scale, in harmonic analysis as the groups from which the nondegenerate series of unitary representations are induced, and in linear algebra
as automorphism groups of singular bilinear and hermitian forms. So they are
worth studying.

0.1. Earlier work on the problem.

Some years ago, Wigner [25] found
all the irreducible unitary representations of the Poincaré group. A number
of authors noted the extent to which Wigner's considerations extend without
change. Quite recently Wigner's result was generalized to representations of
certain maximal parabolic subgroups of indefinite orthogonal, unitary and
symplectic groups [30].

Let F be one of the four finite dimensional real division algebras.
Given integers $p,q \geqslant 0$ we write $F^{p,q}$ for the (right) vector space F^{p+q}
with hermitian form $h(x,y) = \sum\limits_{1 \leqslant \ell \leqslant p} x^{\ell} \overline{y^{\ell}} - \sum\limits_{1 \leqslant m \leqslant q} x^{p+m} \overline{y^{p+m}}$. That gives us

$$U(p,q;F): \text{ the unitary group of } F^{p,q}.$$

Received by the Editors: March 14, 1975.

Research partially supported by NSF Grant MPS 74-01477.

1

$U(p,q;R)$ is the indefinite orthogonal group $O(p,q)$, $U(p,q;C)$ is the usual indefinite unitary group $U(p,q)$, and $U(p,q;Q)$ is the indefinite symplectic (quaternion-unitary) group $Sp(p,q)$. The *Lorentz group* is $O(1,3) = U(1,3;R)$, and $O(2,4) = U(2,4;R)$ is often referred to as the *conformal group*.

In this setting we have a nilpotent group, which is a sort of Heisenberg group, given by

$$N_{p,q,F}: \text{Im } F + F^{p,q} \quad \text{with} \quad (w,z)(w',z') = (w+w +\text{Im } h(z,z'), z+z')$$

where Im denotes imaginary *component*. $U(p,q;F)$ acts by automorphisms on $N_{p,q,F}$ through its action on $F^{p,q}$, and that gives a semidirect product group

$$G_{p,q,F} = N_{p,q,F} \cdot U(p,q;F).$$

Of course Im $R = 0$, so $N_{p,q,R}$ is the additive group $R^{p,q}$, and $G_{p,q,R}$ is the inhomogeneous indefinite orthogonal group $R^{p,q} \cdot O(p,q)$. The *Poincaré group* is the inhomogeneous Lorentz group $G_{1,3,R} = R^{1,3} \cdot O(1,3)$.

The nonzero elements of F also act by automorphisms on $N_{p,q,F}$, by $a: (w,z) \to (aw\bar{a}, az)$. Writing F^* for the group they generate, we have a larger semidirect product

$$P_{p,q,F} = N_{p,q,F} \cdot \{U(p,q;F) \times F^*\} = G_{p,q,F} \cdot F^* .$$

$P_{1,3,R} = R^{1,3} \cdot \{O(1,3) \times R^*\}$ is the *Poincaré group with scale*. If $F \neq$ Cayley numbers, it turns out ([30, §3], or see §2 below) that $P_{p,q,F}$ sits naturally as a maximal parabolic subgroup of $U(p+1,q+1;F)$. In particular, the Poincaré group with scale is a maximal parabolic subgroup of the conformal group. {This has some consequences for cosomology and for charge conjugation

([19], [20], [22]).}

Wigner [25] wrote out the irreducible unitary representation of the Poincare group $G_{1,3,R}$ using a technique now generally known as the _Mackey little-group method_. His procedure goes through without serious change for all the $G_{p,q,R} = R^{p,q} \cdot O(p,q)$, as follows. Every $v \in R^{p,q}$ specifies a unitary character $\chi_v(z) = e^{ih(v,z)}$ on $R^{p,q}$. χ_v has $O(p,q)$-stabilizer $L_v = \{g \in O(p,q): gv = v\}$, and χ_v extends to $R^{p,q} \cdot L_v$ by $\tilde{\chi}_v(z,g) = \chi_v(z)$. If γ is an irreducible unitary representation of L_v one has an induced unitary representation,

$$\pi_{v,\gamma} = \text{Ind}_{R^{p,q} \cdot L_v \uparrow G_{p,q;R}} (\tilde{\chi}_v \otimes \gamma) \in \hat{G}_{p,q,R} .$$

Here v influences π_v, only to the extent of its $O(p,q)$-orbit, which is characterized by the "mass" $-h(v,v)$, and so there are just four cases:

    (i) $v = 0$ and $L_v = O(p,q)$,

    (ii) $h(v,v) > 0$ and $L_v \cong O(p-1,q)$,

    (iii) $h(v,v) < 0$ and $L_v \cong O(p,q-1)$, and

    (iv) $v \neq 0$ but $h(v,v) = 0$, and [30, §3] $L_v \cong G_{p-1,q-1,R}$.

Now the set $\hat{G}_{p,q,R}$ of all equivalence classes of irreducible unitary representations of $G_{p,q,R}$ is described, in several steps, in terms of the $O(r,s)^{\wedge}$ with $0 \leq r \leq p$ and $0 \leq s \leq q$. This is all quite standard for $p = 0$ and for $p = 1$.

This recursion procedure goes through also for the $G_{p,q,F}$ with F complex or quaternionic [30, §§3,4 and 5]. {The Cayley number case has its own peculiarities [30, §8], which do not concern us here.} $R^{p,q}$ is replaced by the generalized Heisenberg group $N_{p,q,F}$, and $\hat{N}_{p,q,F}$ consists (a) of the unitary characters

$$\chi_v \colon\ (w,z) \to e^{i\ \mathrm{Re}\ h(v,z)},\ v \in F^{p,q},$$

and (b) of certain infinite dimensional representations η_λ. Here λ is a nonzero linear functional on $\mathrm{Im}\ F$, and η_λ is characterized up to unitary equivalence by $\eta_\lambda(w,z) = e^{i\lambda(w)}\eta_\lambda(0,z)$. χ_v has $U(p,q;F)$-stabilizer $L_v = \{g \in U(p,q;F)\colon gv = v\}$ as before:

   (i) $v = 0$ and $L_v = U(p,q;F)$,

   (ii) $h(v,v) > 0$ and $L_v \cong U(p-1,q;F)$,

   (iii) $h(v,v) < 0$ and $L_v \cong U(p,q-1;F)$, or

   (iv) $v \neq 0$ but $h(v,v) = 0$, and $L_v \cong G_{p-1,q-1,F}$.

χ_v extends to a character $\tilde{\chi}_v(w,z;g) = \chi_v(w,z)$ on $N_{p,q,F}\cdot L_v$, and we get induced representations $\pi_{v,\gamma}, \gamma \in \hat{L}_v$, of $G_{p,q,F}$. The matter is more subtle with the η_λ, which are stabilized by all of $U(p,q;F)$. In principle one can calculate the Mackey obstruction to extension, but in practice it is better to realize η_λ in a geometric setting on which $U(p,q;F)$ acts naturally [30, §4]. That gives an extension $\tilde{\eta}_\lambda$ of η_λ to $G_{p,q,F}$, resulting in representations $\tilde{\eta}_\lambda \otimes \gamma,\ \gamma \in U(p,q;F)^{\wedge}$. Now $\hat{G}_{p,q,F}$ consists of the classes $[\pi_{v,\gamma}]$ and $[\tilde{\eta}_\lambda \otimes \gamma]$, and so is described, in several steps, in terms of the $U(r,s;F)^{\wedge}$, $0 \leqslant r \leqslant p$ and $0 \leqslant s \leqslant q$.

   Passage from $\hat{G}_{p,q,F}$ to $\hat{P}_{p,q,F}$ is a routine application of induced representations.

   0.2. Classical Lie groups and their parabolic subgroups. We turn to the classical groups and their parabolic subgroups, describing them in a way that suggests an extension of the procedure mentioned in §0.1.

   F will denote a field R (real), C (complex) or Q (quaternion).

The *general linear group* GL(n;F) consists of all invertible F-linear transformations of F^n.

The *special linear group* SL(n;F) is the derived group (commutator sub-group) of GL(n;F). In the real and complex cases it is {g ∈ GL(n;F): det(g) = 1} where det denotes determinant.

The F-*unitary groups* are the U(p,q;F) described in §0.1: the *indefinite orthogonal groups* O(p,q) = U(p,q;R), the *indefinite unitary groups* U(p,q) = U(p,q;C) and the *indefinite symplectic groups* Sp(p,q) = U(p,q;Q). One also has the *indefinite special orthogonal groups* SO(p,q) = {g ∈ O(p,q): det(g) = 1} and the *indefinite special unitary groups* SU(p,q) = {g ∈ U(p,q): det(g) = 1}. These are the groups defined by F-hermitian forms.

The real and complex *symplectic groups* are defined by antisymmetric bilinear forms:

$$Sp(n;F) = \{g \in GL(2n;F): g \text{ preserves } \sum_{1 \leq \ell \leq n} (x^\ell y^{n+\ell} - x^{n+\ell} y^\ell)\},$$

F = R or C. There is no quaternion symplectic group: one does not have bilinear forms over a noncommutative field.

The *complex orthogonal group* is defined by a symmetric bilinear form:

$$O(n;C) = \{g \in GL(n;C): g \text{ preserves } \sum_{1 \leq \ell \leq n} x^\ell y^\ell\}.$$

The real orthogonal groups were already given as R-unitary groups, and there is no quaternion orthogonal group.

Finally one has a group (which has no standard name) defined by an antihermitian form:

$$SO^*(2n) = \{g \in GL(n;Q): \ g \text{ preserves } \kappa(x,y) = \sum_{1 \leqslant \ell \leqslant n} x^\ell i \ \bar{y}^\ell\}.$$

Antihermitian forms on C^n define indefinite unitary groups, and on R^{2n} they just define real symplectic groups.

The _complex classical groups_ are $GL(n;C)$, $SL(n;C)$, $O(n;C)$ and $Sp(n;C)$. The _real classical groups_ are their real forms, which are

$GL(n;C)$: $U(p,q)$ for p+q=n, $GL(n;R)$, and $GL(\frac{1}{2}n;Q)$ for n even.

$SL(n;C)$: $SU(p,q)$ for p+q=n, $SL(n;R)$, and $SL(\frac{1}{2}n;Q)$ for n even.

$O(n;C)$: $O(p,q)$ for p+q=n and $SO^*(n)$ for n even.

$Sp(n;C)$: $Sp(p,q)$ for p+q=n and $Sp(n;R)$.

We have just described the classical groups in the following way. We have a vector space X over F and a form $\varphi: X \times X \to F$ which is symmetric bilinear, antisymmetric bilinear, hermitian, or antihermitian. Either $\varphi = 0$ or φ is nondegenerate. The classical group is $U(X,\varphi) = \{g: X \to X$ linear: g preserves $\varphi\}$ or $SU(X,\varphi) = U(X,\varphi) \cap SL(X)$. The parabolic subgroups (see [23], [3], [28] or [29] for the general theory) of the classical groups are the

$$P_{E_1,\ldots,E_k} = \{g: gE_\ell = E_\ell \quad \text{for } 1 \leq \ell \leq k\}$$

where $0 \neq E_1 \subset \ldots \subset E_k \subset X$ is a sequence of isotropic $(\varphi(E_\ell,E_\ell) = 0)$ subspaces. In each case, the classical group is transitive on the collection of all such isotropic subspace sequences with a fixed dimension sequence $0 < \dim E_1 < \ldots < \dim E_k < \dim X$, and so $P_{E_1,\ldots,E_k}$ is determined up to conjugacy by the $\dim E_\ell$. In particular, the maximal parabolic subgroups are the

$P_E = \{g: gE = E\}$, E proper isotropic subspace of X,

and $\dim E$ specifies the conjugacy class of P_E.

In each case, P_E turns out to have the following structure. There is a totally isotropic subspace $E' \subset X$ such that

if $\varphi = 0$: then $E' = 0$

if $\varphi \neq 0$: $\dim E' = \dim E$ and φ is nondegenerate on $E + E'$.

That gives a splitting $X = V \oplus W$ where

if $\varphi = 0$: $V = E + E' = E$ and W is any complement,

if $\varphi \neq 0$: $V = E + E'$ and $W = V^{\perp}$ relative to φ.

Let $\mathbf{u}$ denote the Lie algebra of our classical group. Then the unipotent radical P_E is the analytic subgroup N with Lie algebra $n = n_1 + n_2$ where

$$n_1 = \text{1-root space} = \{\xi \in \mathbf{u} : \xi E' \subset W, \xi W \subset E \text{ and } \xi E = 0\},$$
$$n_2 = \text{2-root space} = \{\xi \in \mathbf{u} : \xi E' \subset E \text{ and } \xi(E + W) = 0\}.$$

View $U(W, \varphi|_W) \subset U(X, \varphi)$ as all elements that act by the identity on V, and view $GL(E)$ as all elements $g \in U(X, \varphi)$ such that $gE = E$, $gE' = E'$ and $g|_W = $ identity. Denote

$$G_E = \{g \in P_E: g|_E = \text{identity}\}, \text{ kernel of action of } P_E \text{ on } E.$$

If the classical group is $U(X, \varphi)$, then

$$P_E = N \cdot \{U(W,\varphi|_W) \times GL(E)\} \quad \text{and} \quad G_E = N \cdot U(W,\varphi|_W).$$

If the classical group is $SU(X,\varphi)$, then

$$P_E = N \cdot \{SU(W,\varphi|_W) \times GL(E)\} \quad \text{and} \quad G_E = N \cdot SU(W,\varphi|_W).$$

0.3. <u>Representations of the maximal parabolic subgroups</u>. Keeping in mind the picture sketched in §0.2, we indicate the modification of the procedure of §0.1 that gives the irreducible unitary representations of the maximal parabolic subgroups of the classical groups.

The unipotent radical N of P_E is abelian or 2-step nilpotent, so it is more or less similar to the Heisenberg group. We write down a nondegenerate $U(W,\varphi|_W)$-invariant real inner product $< \ , \ >$ on the Lie algebra $\mathfrak{n}$ of N. That is the point at which things start to diverge for the various series of classical groups. Then the real dual space $\mathfrak{n}^*$ is identified with the $f_\xi(\eta) = <\xi,\eta>, \xi \in \mathfrak{n}$, the orbit structure of $\mathfrak{n}^*$ under the (co-adjoint) action of N is calculated, and we describe $\hat{N}$ by a straightforward application of the Kirillov Theorey ([4]; or see [2] or [18]).

If $\xi \in \mathfrak{n}$ let $[\pi_\xi] \in \hat{N}$ denote the unitary representation class associated to the orbit $Ad^*(N) \cdot f_\xi$. In each case, it turns out that the $U(W,\varphi|_W)$-stabilizer of $[\pi_\xi]$ is a group, say L_ξ, of the same sort as G_E, but with smaller matrices. More precisely, there is a subspace $S_\xi \subset W$ such that $L_\xi = \{g \in U(W,\varphi|_W): gx = x \text{ all } x \in S_\xi\}$. S_ξ is the sum of the φ-isotropic space $S_\xi \cap S_\xi^\perp$ and a φ-nondegenerate complement T_ξ, and so $L_\xi = \{g \in U(W \cap T_\xi^\perp, \varphi|_{W \cap T_\xi^\perp}): gx = x \text{ for } x \in S_\xi \cap S_\xi^\perp\}$, or the corresponding group for SU, which is a "baby" version of G_E.

π_ξ usually extends to a unitary representation $\tilde{\pi}_\xi$ of $N \cdot L_\xi$ on the same Hilbert space. Roughly speaking, this is a matter of the possibility of realizing $[\pi_\xi]$ in a geometric setting on which L_ξ acts naturally. But in the case of the real and complex symplectic groups we are forced to calculate the Mackey obstruction to extension. That obstruction can be nontrivial for $Sp(n;R)$, but it is always killed by passage to the 2-sheeted covering group $Mp(n;R)$, the "metaplectic group." This technicality aside, now $\hat{G}_E = \{[\text{Ind}_{N \cdot L_\xi \uparrow G_E}(\tilde{\pi}_\xi \otimes \gamma)] : \xi \in \mathfrak{n} \text{ and } [\gamma] \in \hat{L}_\xi\}$. As L_ξ is a baby version of G_E, we know $\hat{L}_\xi$ by recursion on the size of the matrices. Thus $\hat{G}_E$ is completely described in terms of the $U(Y, \varphi|_Y)\hat{}$ or $SU(Y, \varphi|_Y)\hat{}$ where $Y \subsetneq X$ and where $\varphi|_Y$ is nondegenerate when φ is nondegenerate.

Let $\xi \in \mathfrak{n}$, $[\gamma] \in \hat{L}_\xi$, and $[\pi_{\xi,\gamma}] \in \hat{G}_E$ the class induced from $[\tilde{\pi}_\xi \otimes \gamma]$. The $GL(E)$-stabilizer of $[\pi_{\xi,\gamma}]$ does not depend on γ, so denote it by M_ξ. Then M_ξ always turns out to be a quotient group of a parabolic subgroup of a classical group, but *usually that classical group is not of the type with which we started*. For example, if we start with $O(p,q)$, the M_ξ force us to consider the $Sp(m;R)$; if we start with $Sp(p,q)$, the M_ξ force us to consider the $SO^*(2m)$; if we start with $Sp(m;C)$, the M_ξ force us to consider the $O(d;C)$. This problem did not occur in the earlier work [30] mentioned in §0.1 because there $\dim E = 1$. At any rate, since we are considering all the classical groups, we are able to thread our way through the tangle and describe $\hat{M}_\xi$. Furthermore, $\pi_{\xi,\gamma}$ always turns out to extend to a unitary representation $\tilde{\pi}_{\xi,\gamma}$ of $G_E \cdot M_\xi$ on the same Hilbert space, and so

$$\hat{P}_E = \{[\text{Ind}_{G_E \cdot M_\xi \uparrow P_E}(\tilde{\pi}_{\xi,\gamma} \otimes \mu)] : \xi \in \mathfrak{n}, \ [\gamma] \in \hat{L}_\xi, \ [\mu] \in \hat{M}_\xi\}$$

gives a complete description of the irreducible unitary representation classes of P_E in terms of the unitary duals of smaller classical groups.

0.4. Parabolic subgroups in general. Let U be a reductive Lie group, P a parabolic subgroup, and N the unipotent radical of P. If N is not 2-step nilpotent, $\hat{N}$ becomes somewhat complicated, and the description of $\hat{P}$

cannot be carried out by our rather simple methods. Those methods are based
on a detailed knowledge of the representatives of the classes in $\hat{N}$ and
explicit descriptions of their stabilizers. Here it is instructive to try
to write out $\hat{P}$ where P is a minimal parabolic subgroup in one of the
simple groups

$$SU(2,m), \quad SO(2,m), \; Sp(2,m), \; SO^*(8 \text{ or } 10), \; SO(5;\mathbb{C}), \; E_{6,C_4}$$

with restricted root system of type B_2.

At this point one asks which parabolic subgroups have unipotent radical
which is abelian or 2-step nilpotent. The problem reduces immediately to
the case of parabolic subgroups $P \subsetneq U$ where U is noncompact and simple.
There, let $P = N \cdot P_r$ where N is the unipotent radical and P_r is a
reductive complement. Lower case German letters denote Lie algebras. Let
$\mathfrak{z}$ be the center of $\mathfrak{p}_r$. If $\mathfrak{u}$ is complex, let $\Sigma^+_{\mathfrak{z}}$ denote positive $\mathfrak{z}$-root
system on $\mathfrak{u}$ such that $\mathfrak{n}$ is the sum of the positive $\mathfrak{z}$-root spaces. If $\mathfrak{u}$
is absolutely simple let $\Sigma^+_{\mathfrak{z}}$ denote the positive $\mathfrak{z}_C$-root system on $\mathfrak{u}_C$
such that $\mathfrak{n}_C$ is the sum of the positive $\mathfrak{z}_C$-root spaces. Using [26,
Theorem 8.13.3] one can see that

$$N \text{ is abelian} \qquad \Leftrightarrow \Sigma^+_{\mathfrak{z}} \text{ has form } \{\alpha\}, \text{ and}$$

$$N \text{ is 2-step nilpotent} \quad \Leftrightarrow \Sigma^+_{\mathfrak{z}} \text{ has form } \{\alpha, 2\alpha\} \text{ or } \{\alpha, \beta, \alpha+\beta\}.$$

One can list all such parabolic subgroups [31]. If $\Sigma^+_{\mathfrak{z}}$ is of the form
or $\{\alpha, 2\alpha\}$ then P is maximal. If $\Sigma^+_{\mathfrak{z}}$ is of the form $\{\alpha, \beta, \alpha+\beta\}$, then
P is maximal if and only if α and β have the same restriction to the
R-split component of $\mathfrak{z}$. Both types of maximal parabolic subgroups occur
in this paper.

We will return to the general case of a parabolic subgroup, whose unipotent radical is abelian or 2-step nilpotent, in [31]. There we make essential use of the specific results of this Memoir.

0.5. Notes to the reader. We do assume some familiarity with matrix groups and basic representation theory, but we do not expect the reader to be acquainted with the theory of semisimple Lie groups and parabolic subgroups. Everything that we need from the theory of induced representations is summarized in the Appendix, and for the most part (i.e. except when dealing with $Sp(n;R)$) the contents of §§A.I and A.II will suffice.

For reasons of clarity, the method sketched in §§0.2 and 0.3 is not carried out in general *ab initio*. Instead, it is carried out in detail in Part II, for the groups $U(p,q;F)$, and modifications for the other groups are indicated as needed. The point is that there is too much variation of technical detail between the various series of classical groups.

Our results are written out for $GL(n;F)$, $U(p,q;F)$, $Sp(n;F)$ and $Mp(n;F)$, $O(n;C)$, and $SO^*(2m)$. The reader will have no difficulty in transposing them to $GL'(n;F)$ and $SL(n;F)$, $SU(p,q;F)$, and $SO(n;C)$. See [29] for the method of passing to covering groups.

$SO^*(2m)$ is mainly viewed as $O(2m;C) \cap U(m,m)$, inside $GL(2m;C)$, in Part IV, but its realization as the subgroup of $GL(m;Q)$ defined by a skew-hermitian form is inescapable.

P is reserved for parabolic subgroups and groups that turn out to be isomorphic to parabolics, G for the subgroups $(g|_E = \text{identity})$ of parabolics, N for their unipotent radicals. We usually write (D,Z) for an element of $\mathfrak{n} = \mathfrak{n}_2 + \mathfrak{n}_1$, $f_{D,Z}$ for the corresponding linear functional, $[\pi_{D,Z}]$ for the class in $\hat{N}$ associated to the orbit of $f_{D,Z}$. Expressing $G = N \cdot U(W,\varphi|_W)$ and $P = G \cdot GL(E)$, we write $L_{D,Z}$ for the $U(W,\varphi|_W)$-stabilizer of $[\pi_{D,Z}]$. Given $[\gamma] \in \hat{L}_{D,Z}$ we write $[\pi_{D,Z,\gamma}]$ for the class in $\hat{G}$ induced from $\tilde{\pi}_{D,Z} \otimes \gamma$ where $\tilde{\pi}_{D,Z}$ is an extension of $\pi_{D,Z}$ to $N \cdot L_{D,Z}$.

Joseph A. Wolf

The $GL(E)$-stabilizer of $[\pi_{D,Z,\gamma}]$ is independent of $[\gamma]$ and is denoted $M_{D,Z}$. Given $[\mu] \in \hat{M}_{D,Z}$, we write $[\pi_{D,Z,\gamma,\mu}]$ for the class in $\hat{P}$ induced from $\tilde{\pi}_{D,Z,\gamma} \otimes \mu$. Finally, J is used for the abstract models of the $M_{D,Z}$.

Part I. Linear Groups

§1. Parabolic Subgroups of General Linear Groups.

In the section we write out the structure and representation theory for maximal parabolic subgroups of real, complex and quaternionic general linear groups. Although straightforward, this is of some intrinsic interest. More to the point, it illustrates the method for parabolic subgroups of unitary groups in a relatively simple setting, and we need the result in the sequel for application to parabolic subgroups of the other classical groups.

F will denote a real division algebra R (real numbers), C (complex numbers) or Q (quaternions). F^n is the right vector space of n-tuples from F. Since the scalars act on the right, the linear transformations act on the left, and one should think of the elements of F^n as column vectors. The invertible F-linear transformations of F^n form a reductive real Lie group

(1.1)   GL(n;F): general linear group of degree n over F.

We view GL(n,F) as consisting of all invertible n × n matrices over F, acting on F^n by matrix multiplication.

A _flag_ in F^n is an increasing sequence $0 \neq E_1 \subsetneq E_2 \subsetneq \cdots \subsetneq E_k \subsetneq F^n$ of F-linear subspaces. The associated subgroups

(1.2a) $P_{E_1,\ldots,E_k} = \{g \in GL(n;F): gE_\ell = E_\ell \text{ for } 1 \leq \ell \leq k\}$

are the _parabolic subgroups_ of GL(n;F). The reader not acquainted with parabolic subgroups in general may take this as the definition. If $0 \neq E_1' \subsetneq \cdots \subsetneq E_k' \subsetneq F^n$ is another flag, and if $\dim E_\ell' = \dim E_\ell$ for $1 \leq \ell \leq k$, then one has an element $g \in GL(n;F)$ with each $gE_\ell' = E_\ell$, and so

$$g \cdot P_{E_1', \ldots, E_\ell'} \cdot g^{-1} = P_{E_1, \ldots, E_\ell} \cdot$$

Thus the conjugacy class of the parabolic subgroup (1.2a) depends only on the
sequence $0 < \dim E_1 < \ldots < \dim E_k < n$ of dimensions, and the space of all
flags with that dimension sequence is a real analytic manifold
$GL(n;F)/P_{E_1, \ldots, E_k}$, called a *flag manifold*. Now let us note that the
maximal parabolic subgroups of $GL(n;F)$ are the

$$(1.2b) \qquad\qquad P_E = \{g \in GL(n;F): gE = E\}$$

where E is a linear subspace with $0 < \dim E < n$. Thus $GL(n;F)$ has
precisely n-1 conjugacy classes of maximal parabolic subgroups.

Let a and b be positive integers and let $F^{a \times b}$ denote the real
vector space of all $a \times b$ matrices over F. $GL(a;F) \times GL(b;F)$ acts on
$F^{a \times b}$ as a group of real linear transformations by the formula
$(A,B): X \to AXB^{-1}$. That gives us the semidirect product group

$$(1.3a) \qquad\qquad L_{a,b}(F) = F^{a \times b} \cdot \{GL(a;F) \times GL(b;F)\}$$

whose group law is

$$(1.3b) \qquad\qquad (X;A,B)(X';A',B') = (X + AX'B^{-1}; AA', BB').$$

We will also have use for the subgroup given by $A = I$:

$$(1.3c) \qquad K_{a,b}(F) = F^{a \times b} \cdot GL(b;F) \quad \text{with} \quad (X;B)(X';B') = (X + X'B^{-1}; BB').$$

1.4. Lemma. _Let_ E _be a subspace of dimension_ a _in_ F^{a+b} . _Then there is a Lie group isomorphism_

$$\zeta: L_{a,b}(F) \cong P_E = \{g \in GL(a+b;F): gE = E\}$$

such that

$$\zeta: K_{a,b}(F) \cong \{g \in GL(a+b;F): gx = x \text{ _for all_ } x \in E\}.$$

Proof: Applying an element of $GL(a+b;F)$ we may assume that E is spanned by the first a vectors in a standard basis of F^{a+b}. Then P_E consists of all matrices $\begin{bmatrix} A & Z \\ 0 & B \end{bmatrix}$ over F where A is a × a invertible and B is b × b invertible, and $\zeta(X;A,B) = \begin{bmatrix} A & XB \\ 0 & B \end{bmatrix}$ is the required isomorphism.   _q.e.d._

$F^{a \times b}$ carries a positive definite real inner product
$\langle X,Y \rangle = $ Re trace XY^*. Here Y^* denotes the conjugate (of F over R) transpose of Y, trace means the sum of the diagonal matrix entries, and Re means real part. Thus the unitary characters on $F^{a \times b}$ are the

$$(1.5) \qquad\qquad \chi_Y: X \to e^{i \text{ Re trace } XY^*} = e^{i\langle X,Y \rangle}.$$

We calculate the action of $GL(a;F) \times GL(b;F)$ on the χ_Y:

$$\{(A,B)\chi_Y\}(X) = \chi_Y\{(A,B)^{-1}X\} = \chi_Y(A^{-1}XB)$$

$$= e^{i \text{ Re trace } A^{-1}XBY^*} = e^{i \text{ Re trace } XBY^*A^{-1}} = \chi_{(A^{-1})^*YB^*}(X),$$

that is,

(1.6) $(A,B)\chi_Y = \chi_{Y'}$, with $Y' = (A^{-1})^* Y B^*$.

In particular, the stabilizer of χ_Y in $GL(a;F) \times GL(b;F)$ is the subgroup given by $BY^* A^{-1} = Y^*$.

   Mackey's little group method is described in the Appendix. Here we need a special case, which we recall from §A.II. Write $\widehat{}$ for unitary dual, the set of all equivalence classes of irreducible unitary representations, with its usual Borel structure. If $G = N \cdot U$ semidirect product of locally compact groups, then the little-group method describes $\widehat{G}$ in terms of $\widehat{N}$ and the unitary duals of certain subgroups of U, as follows. Let $[\eta] \in \widehat{N}$, denote its stabilizer

(1.7a) $G_\eta = N \cdot U_\eta = \{g \in G \colon n \to \eta(g^{-1}ng)$ is equivalent to $\eta\}$,

and consider the "extensions"

(1.7b) $\mathscr{E}(\eta) = \{[\psi] \in \widehat{G}_\eta \colon \psi|_N$ is equivalent to a multiple of $\eta\}$.

If all these groups are type I and if there is a Borel section to the action of G on $\widehat{N}$, then $\widehat{G}$ consists of the classes unitarily induced from the various $\psi \in \mathscr{E}(\eta)$:

(1.7c) $\widehat{G} = \{[\mathrm{Ind}_{G_\eta \uparrow G}(\psi)] \colon [\eta] \in \widehat{N}$ and $[\psi] \in \mathscr{E}(\eta)\}$.

These conditions are automatic for our groups, which always will be real linear algebraic groups with G acting analytically on $\widehat{N}$. In our case matters will be further simplified because, given $[\eta] \in \widehat{N}$, we will find $[\tilde{\eta}] \in \widehat{G}_\eta$ with $\tilde{\eta}|_N = \eta$. That implies

(1.8a) $\mathscr{E}(\eta) = \{[\tilde{\eta} \otimes \mu] \colon [\mu] \in \widehat{U}_\eta$ extended by $\mu(n,u) = \mu(u)\}$

and so we will have (see (A.13′) in the Appendix)

$$(1.8\text{b}) \qquad \hat{G} = \{[\text{Ind}_{G_\eta \uparrow G}(\tilde{\eta} \otimes \mu)] : [\eta] \in \hat{N} \text{ and } [\mu] \in \hat{U}_\eta\} \ .$$

We first apply the little-group method to $K_{a,b}(F) = F^{a \times b} \cdot GL(b;F)$ and then apply it again to $L_{a,b}(F) = K_{a,b}(F) \cdot GL(a;F)$. In view of Lemma 1.4 this gives the unitary duals of the maximal parabolic subgroups of the real, complex and quaternionic general linear groups.

1.9. Lemma. _Let_ $Y \in F^{a \times b}$ _and let_ E_Y _denote the subspace of_ F^b _spanned by the columns of_ Y^*. _Then_

$$S_Y = \{B \in GL(b;F) : By = y \ \text{\textit{for all}} \ y \in E_Y\}$$

is the $GL(b;F)$-_stabilizer of_ χ_Y , _and_ $F^{a \times b} \cdot S_Y$ _is the stabilizer of_ χ_Y _in_ $K_{a,b}(F)$. _If_ Y _has rank_ c, _that is_ $c = \dim E_Y$, _then_ $S_Y \cong K_{c,b-c}(F)$.

Proof: From (1.6) the $GL(b;F)$-stabilizer of χ_Y is given by $BY^* = Y^*$, which is equivalent to $By = y$ for all y in the column span E_Y of Y^*, which in turn is the condition $B \in S_Y$. The second assertion follows, and $S_Y \cong K_{c,b-c}(F)$ by Lemma 1.4.   _q.e.d._

1.10. Lemma. _If_ $Y \in F^{a \times b}$ _then_ $\tilde{\chi}_Y(X;B) = \chi_Y(X)$ _defines a unitary character on_ $F^{a \times b} \cdot S_Y$.

Proof. Let $X, X' \in F^{a \times b}$ and $B, B' \in S_Y$. Then $(X;B)(X';B') = (X + X'B^{-1}, BB')$ and we calculate $\langle X + X'B^{-1}, Y \rangle = \langle X, Y \rangle + \langle X'B^{-1}, Y \rangle = \langle X, Y \rangle + \langle X', Y \rangle$ because $B^{-1}Y^* = Y^*$.   _q.e.d._

By the little-group method, $K_{a,b}(F)^\wedge$ consists of the unitarily induced classes

(1.11a) $[\kappa_{Y,\sigma}] = [\text{Ind}_{F^{a\times b}\cdot S_Y \uparrow K_{a,b}(F)} (\tilde{\chi}_Y \otimes \sigma)]$

where

(1.11b) $Y \in F^{a\times b}$ and $[\sigma] \in \hat{S}_Y$ extended by $\sigma(X;B) = \sigma(B)$.

The equivalence criterion for representations obtained by the little group method -- see (A.10) in the Appendix -- tells us that $[\kappa_{Y,\sigma}] = [\kappa_{Y',\sigma'}]$ if, and only if, there exists $B \in GL(b;F)$ such that

(1.11c) $Y' = YB^*$ and $B' \to \sigma(BB'B^{-1})$ is equivalent to σ'.

Note that $Y' = YB^*$ if and only if B carries the columns of Y^* to be corresponding columns of Y'^*. Now we use Lemma 1.9 to reformulate (1.11) as

   1.12. Theorem. $K_{a,b}(F)\hat{}$ _is the disjoint union of_ $1 + \min(a,b)$ _subsets_ $K_{a,b}(F)\hat{}_c$ _as follows. Let_ $E^{c\times b}(F)$ _denote the set of all_ $c \times b$ _echelon matrices_

$$\begin{bmatrix} I & * & 0 & * & 0 & * & & & 0 & * & 0 & * \\ & & I & * & 0 & * & & & & \vdots & & \\ & & & & I & * & & & & \vdots & & \\ & & & & & & \ddots & & 0 & * & 0 & * \\ & 0 & & & & & & \ddots & I & * & 0 & * \\ & & & & & & & & & & I & * \end{bmatrix}$$

of rank c _with entries in_ F. _Then_ $K_{a,b}(F)\hat{}_c$ _is parametrized by_ $E^{c\times b}(F) \times K_{c,b-c}(F)\hat{}$ _under_

$$(Z,[\sigma]) \leftrightarrow [\kappa_{Y,\sigma}] \quad with \quad Y = \begin{pmatrix} Z \\ 0 \end{pmatrix} \in F^{a\times b}$$

where .

$$S_Y \ \textit{is identified with its isomorph} \ K_{c,b-c}(F)$$

Proof. It only remains to be shown that a matrix $Y' \in F^{a \times b}$ of rank c is of the form YB^* with $B \in GL(b;F)$ and $Y = \begin{pmatrix} Z \\ 0 \end{pmatrix}$, for a unique matrix $Z \in E^{c \times b}(F)$. That is just the ordinary elementary row-echelon normal form.

$$q.e.d.$$

1.13. Remark. Theorem 1.12 reduces the explicit determination of $K_{a,b}(F)^{\wedge}$ to that of $GL(\ell;F)^{\wedge}$ for $\ell \leq b$. In effect, the case $c = 0$ gives $E^{0 \times b}(F) = \{0\}$ and $K_{0,b}(F) = GL(b;F)$, so $K_{a,b}(F)^{\wedge}_0$ consists of the classes $[\sigma] \in GL(b;F)^{\wedge}$ lifted to $K_{a,b}(F)$. If $c > 0$, then $K_{c,b-c}(F)^{\wedge}$ is already given by Theorem 1.12, recursion and the $GL(\ell;F)^{\wedge}$ for $\ell \leq b-c$. The representation theory of $GL(\ell;F)^{\wedge}$ is more or less known, for example through the work of Gelfand's Moscow school and also through the work of Harish-Chandra.

1.14. Lemma. *Let* $[\kappa_{Y,\sigma}] \in K_{a,b}(F)^{\wedge}$ *as in* (1.11). *Its* $GL(a;F)$-*stabilizer is*

$$T_Y = \{A \in GL(a;F): A^* \cdot Y \in Y \cdot GL(b;F)\}$$

and $K_{a,b}(F) \cdot T_Y$ *is its* $L_{a,b}(F)$-*stabilizer. If* E^*_Y *denotes the subspace of* F^a *spanned by the columns of* Y *then*

$$T_Y = \{A \in GL(a;F): A^*(E^*_Y) = E^*_Y\}.$$

In particular, if $c = $ rank Y, *that is* dim $E^*_Y = c$, *then* $T_Y \cong L_{c,a-c}(F)$.

Proof. Since $GL(a;F)$ centralizes $GL(b;F)$ in $L_{a,b}(F)$, (1.6) gives us $A[\kappa_{Y,\sigma}] = [\kappa_{Y',\sigma'}]$ with $Y' = (A^{-1})^*Y$ and $\sigma' = \sigma$. Thus $A[\kappa_{Y,\sigma}] = [\kappa_{Y,\sigma}]$ if and only if $(A^{-1})^* \cdot Y \in Y \cdot GL(b;F)$, which is equivalent to $A^* \cdot Y \in Y \cdot GL(b;F)$. Now the $GL(a;F)$-stabilizer of $[\kappa_{Y,\sigma}]$ is the subgroup T_Y as asserted.

The action $Y \to YB$ on the right vector space E_Y^* spanned by the columns of Y, is just an arbitrary invertible linear transformation. Thus $A \in T_Y$ precisely when $A^*(E_Y^*) = E_Y^*$, and now Lemma 1.4 gives us $T_Y \cong \{(A^{-1})^* : A \in T_Y\} \cong L_{c,a-c}(F)$.

1.15. Lemma. *If* $[\kappa_{Y,\sigma}] \in K_{a,b}(F)^{\wedge}$ *then there is a class* $[\tilde{\kappa}_{Y,\sigma}] \in \{K_{a,b}(F) \cdot T_Y\}^{\wedge}$ *whose restriction to* $K_{a,b}(F)$ *is* $[\kappa_{Y,\sigma}]$.

Proof. $Q_Y = \{(A,B) \in GL(a;F) \times GL(b;F) : A^*Y = YB^*\}$ is the $GL(a;F) \times GL(b;F)$ stabilizer of χ_Y, by (1.6). As in Lemma 1.10, $\chi_Y'(X;A,B) = \chi_Y(X)$ defines a unitary character on $F^{a \times b} \cdot Q_Y$ that extends χ_Y.

Let $\tilde{S}_Y = \{B \in GL(b;F) : YB^* \in GL(a;F) \cdot Y\}$. Lemma 1.9 and the obvious variation on Lemma 1.14 give an isomorphism of $\tilde{S}_Y$ onto $L_{c,b-c}(F)$ that carries S_Y onto $K_{c,b-c}(F)$. Let U_σ denote the S_Y-stabilizer of $[\sigma]$. If $c = a$, then $c \leqslant \min(a,b)$ forces $a = b = c$, so $S_Y = \{1\}$, σ is trivial, and σ extends to the trivial representation of $\tilde{S}_Y \cong GL(c;F)$. If $c < a$, then, by induction on a, we have a class $[\tilde{\sigma}] \in \hat{U}$ such that $[\tilde{\sigma}|_{S_Y}] = [\sigma]$.

Now $[\tilde{\chi}_Y \otimes \sigma]$ extends to a class $[\gamma_{Y,\sigma}]$ on $G_{Y,\sigma} = F^{a \times b} \cdot \{Q_Y \cap (GL(a;F) \times U_\sigma)\}$ by the formula $\gamma_{Y,\sigma}(X;A,B) = \chi_Y'(X;A,B) \cdot \tilde{\sigma}(B)$. Note that $G_{Y,\sigma}$ is the stabilizer of $[\tilde{\chi}_Y \otimes \sigma]$ in $F^{a \times b} \cdot \{GL(a;F) \times GL(b;F)\} = L_{a,b}(F)$, and that $G_{Y,\sigma} \cdot (1 \times GL(b;F)) = F^{a \times b} \cdot (\tilde{T}_Y \times GL(b;F)) = K_{a,b}(F) \cdot T_Y$. Thus we have an irreducible unitary representation

$$\tilde{\kappa}_{Y,\sigma} = \mathrm{Ind}_{G_{Y,\sigma}\uparrow K_{a,b}(F)\cdot T_Y}(\gamma_{Y,\sigma})$$

of $K_{a,b}(F)\cdot T_Y$. Further, $\dot{Q}_Y \cap (1 \times GL(b;F)) = S_Y$ gives $G_{Y,\sigma} \cap K_{a,b}(F) = F^{a\times b}\cdot S_Y$, so the base space of the induction procedure (fibre bundle formulation--see remark following (A.5)) is

$$K_{a,b}(F)\cdot T_Y/G_{Y,\sigma} \approx K_{a,b}(F)/F^{a\times b}\cdot S_Y.$$

Thus the restriction of $\tilde{\kappa}_{Y,\sigma}$ to $K_{a,b}(F)$ is

$$\mathrm{Ind}_{F^{a\times b}\cdot S_Y\uparrow K_{a,b}(F)}\left(\gamma_{Y,\sigma}\Big|_{F^{a\times b}\cdot S_Y}\right) = \mathrm{Ind}_{F^{a\times b}\cdot S_Y\uparrow K_{a,b}(F)}(\tilde{\chi}_Y \otimes \sigma) = \kappa_{Y,\sigma}$$

as required.   *q.e.d.*

By the little-group method now $L_{a,b}(F)^{\wedge}$ consists of the

(1.16a) $$[\lambda_{Y,\sigma,\tau}] = [\mathrm{Ind}_{K_{a,b}(F)\cdot T_Y\uparrow L_{a,b}(F)}(\tilde{\kappa}_{Y,\sigma} \otimes \tau)]$$

where

(1.16b) $Y \in F^{a\times b}$, $[\sigma] \in \hat{S}_Y$, $[\tau] \in \hat{T}_Y$ extended by $\tau(X;A,b) = \tau(B)$.

Now the equivalence criterion in the little group method, (A.10) in the Appendix, tells us that classes $[\lambda_{Y,\sigma,\tau}] = [\lambda_{Y',\sigma',\tau'}]$ if and only if there exists $(A,B) \in GL(a;F) \times GL(b;F)$ such that

(1.16c) $\begin{cases} Y' = (A^*)^{-1}YB^*, \quad \sigma' \text{ is equivalent to } B' \to \sigma(BB'B^{-1}) \\ \\ \text{and} \quad \tau' \text{ is equivalent to } A' \to \tau(AA'A^{-1}). \end{cases}$

Since $Y' = (A^*)^{-1}YB^*$ precisely when Y and Y' differ by a set of elementary row and column transformations, that is rank Y = rank Y' , we can apply Lemma 1.14 to reformulate our discussion as

    1.17 Lemma. $L_{a,b}(F)^\wedge$ _is the disjoint union of_ $1 + \min(a,b)$ _subsets_ $L_{a,b}(F)^\wedge_c$ _as follows. Let_ c _be an integer._ $0 \leqslant c \leqslant \min(a,b)$. _Then_ $L_{a,b}(F)^\wedge_c$ _is parameterized by_

$$K_{c,b-c}(F)^\wedge \times L_{c,a-c}(F)^\wedge$$

under

$$([\sigma],[\tau]) \leftrightarrow [\lambda_{Y,\sigma,\tau}] \quad with \quad Y = \begin{pmatrix} I_c & 0 \\ 0 & 0 \end{pmatrix}$$

where

$$S_Y \ and \ T_Y \ are \ identified \ with \ K_{c,b-c}(F) \ and \ L_{c,a-c}(F).$$

    As in Remark 1.13, Theorem 1.17 reduces the explicit determination of $L_{a,b}(F)^\wedge$ to that of $GL(\ell;F)^\wedge$ for $1 \leqslant \ell \leqslant \max(a,b)$.

    Recall Lemma 1.4. If P is a maximal parabolic subgroup of $GL(n;F)$, now the unitary dual $\hat{P}$ is given by Theorem 1.17, and the explicit determination of all such $\hat{P}$ is reduced to that of the $GL(\ell;F)^\wedge$ for $\ell < n$.

    Finally we note that the entire procedure goes through without essential changes for the maximal parabolic subgroups of the

$$GL'(n;F) = \{\gamma \in GL(n;F): \ \gamma \ \text{preserves Lebesgue measure on} \ F^n\},$$

which have the property $GL(n;F) \cong GL'(n;F) \times R^+$. It also goes through for the subgroups $SL(n;C)$ and $SL(n;R)$, as well as the larger groups

$$GL(n;F) \times (\text{multiplicative group of } F)$$

which act on F^n by (γ, c): $x \rightarrow \gamma x \bar{c}$.

Part II: Unitary Groups

§2. Parabolic Subgroups of Unitary Groups: Statement of Structure

Let F be a real division algebra R (real numbers), C (complex
numbers) or Q (quaternions). View the space F^n of n-tuples from F as
a right vector space, so scalars act on the right and linear transformations
act on the left. If p and q are non-negative integers with p+q = n,
we have the hermitian vector space

$$(2.1) \quad F^{p,q}: \quad F^n \text{ with hermitian form } h(x,y) = \sum_1^p x^{\ell}\overline{y}^{\ell} - \sum_{p+1}^{p+q} x^m\overline{y}^m$$

where ‾ is conjugation of F over R. The F-linear transformations of F^n
that preserve h form a group

$$(2.2) \qquad\qquad U(p,q;F): \text{ unitary group of } F^{p,q}.$$

Here note that U(p,q;R) is the indefinite orthogonal group O(p,q), that
U(p,q;C) is the ordinary indefinite unitary group U(p,q), and that
U(p,q;Q) is the indefinite unitary symplectic group Sp(p,q). In all cases,
U(p,q;F) is a reductive Lie group of real rank r = min(p,q), and it is
compact just when r = 0, i.e. when h is positive or negative definite.

If V_1 and V_2 are linear subspaces of $F^{p,q}$, and if $t: V_1 \to V_2$
is a linear isomorphism such that h(tx,ty) = h(x,y) for all $x, y \in V_1$,
then Witt's Theorem ensures the existence of an element $g \in U(p,q;F)$ such
that $t = g|V_1$. We will make constant use of this fact.

We will say that a basis $\{c_1,\ldots,e_{p+q}\}$ of $F^{p,q}$ is orthonormal if
$h(e_i,e_j) = 0$ for $i \neq j$, 1 for $1 \leqslant i = j \leqslant p$, -1 for
$p + 1 \leqslant i = j \leqslant p + q$.

A linear subspace $E \subset F^{p,q}$ is called *totally isotropic* if $h(E,E) = 0$. If E_1 and E_2 are totally isotropic subspaces of the same dimension, then Witt's theorem gives $g \in U(p,q;F)$ with $g(E_1) = E_2$. In particular, if E is a totally isotropic subspace of dimension s over F, then $s \leqslant r = \min(p,q)$, and $F^{p,q}$ has an orthonormal basis $\{e_1,\ldots,e_{p+q}\}$ such that $\{e_1 + e_{p+1},\ldots,e_s + e_{p+s}\}$ is a basis of E.

It is easy to verify (or the reader unfamiliar with the theory of real semisimple groups may take as definition) that the parabolic subgroups of $U(p,q;F)$ are the

$$(2.3a) \qquad P_{E_1,\ldots,E_k} = \{g \in U(p,q;F) : g(E_\ell) = E_\ell \quad \text{for} \quad 1 \leqslant \ell \leqslant k\}$$

where $0 \neq E_1 \subsetneq E_2 \subsetneq \cdots \subsetneq E_k$ is an increasing sequence of totally isotropic subspaces of $F^{p,q}$. In particular, the maximal parabolic subgroups of $U(p,q;F)$ are the

$$(2.3b) \qquad P_E = \{g \in U(p,q;F) : g(E) = E\} \ , \quad 0 \neq E \quad \text{totally isotropic.}$$

From Witt's Theorem, conjugacy is just a matter of equality of dimension of the corresponding totally isotropic subspaces. In particular, two maximal parabolic subgroups P_E and $P_{E'}$ are conjugate just when $\dim E = \dim E'$, and so there are r conjugacy classes of maximal parabolic subgroups in $U(p,q;F)$.

We now set about describing the Lie group and real algebraic group structure of the maximal parabolic subgroups P_E, and we begin by defining some nilpotent groups that will appear as their unipotent radicals.

$F^{s \times m}$ denotes the space of $s \times m$ matrices over F, so if $A \in F^{s \times m}$ then its adjoint (conjugate transpose) $A^* \in F^{m \times s}$. If $m = t+u$ we have a hermitian map

$$\mathcal{H}: F^{s\times m} \times F^{s\times m} \to F^{s\times s} \quad \text{by} \quad H((A_0,B_0),\ (A,B)) = A_0A^* - B_0B^*$$

where $A_0,\ A \in F^{s\times t}$ and $B_0,\ B \in F^{s\times u}$. We denote

(2.4a) $F^{s\times(t,u)}: F^{s\times m}$ with the hermitian map $\mathcal{H}$.

Further decompose $F^{s\times s} = \text{Re } F^{s\times s} + \text{Im } F^{s\times s}$, direct sum of real vector spaces, under the projections

(2.4b) $\text{Re } D = \frac{1}{2}(D + D^*)$ and $\text{Im } D = \frac{1}{2}(D - D^*)$,

so the summands are the hermitian and skew hermitian components.

Given integers $s \geq 1$ and $t,\ u \geq 0$ we now define nilpotent Lie groups

(2.5a) $N_{s;t,u}(F) = \text{Im } F^{s\times s} + F^{s\times(t,u)}$

with group composition

(2.5b) $(D_0,Z_0)(D,Z) = (D_0 + D + \text{Im } \mathcal{H}(Z_0,Z),\ Z_0 + Z)$.

The generalized Heisenberg groups of [30] are the case $s = 1$. The ordinary Heisenberg group of dimension $2t + 1$ is the case $s = 1$, $u = 0$ and $F = C$. Notice that $N_{s;t,u}(F)$ has derived group in $\text{Im } F^{s\times s}$ and that $\text{Im } F^{s\times s}$ is central. It follows from a glance at (2.5b) that

(2.6a) $N_{s;t,u}(F)$ is abelian $\Leftrightarrow$ either $s = 1$ and $F = R$, or $t = u = 0$;

and

(2.6b) $N_{s;t,u}(F)$ is abelian, or is 2-step nilpotent with center $\mathrm{Im}\ F^{s\times s}$.

Also observe that $N_{s;t,u}(F)$ is the connected simply connected nilpotent Lie group with Lie algebra

(2.7) $n_{s;t,u}(F) = \mathrm{Im}\ F^{s\times s} + F^{s\times(t,u)}$ with $[(D_0,Z_0),(D,Z)] = (2\ \mathrm{Im}\ \mathcal{H}(Z_0,Z),0)$.

Finally, as in [30, Lemma 2.6], $(D, (A,B)) \to (D, (A,\bar{B}))$ defines an isomorphism of $n_{s;t,u}(F)$ onto $n_{s;t+u,0}(F)$, so the isomorphism class of $N_{s;t,u}(F)$ is specified by s, t+u and F.

Choose an orthonormal basis of $F^{t,u}$ and view the unitary group as a matrix group over F,

$$U(t,u;F) = \left\{ g \in F^{(t+u)\times(t+u)} : g\begin{pmatrix} I_t & 0 \\ 0 & -I_u \end{pmatrix}g^* = \begin{pmatrix} I_t & 0 \\ 0 & -I_u \end{pmatrix} \right\}.$$

Then $\mathcal{H}(Z_0 g^*, Zg^*) = \mathcal{H}(Z_0,Z)$ for $Z_0, Z \in F^{s\times(t,u)}$ and $g \in U(t,u;F)$, so $U(t,u;F)$ acts by automorphisms on $N_{s;t,u}(F)$ by $g(D,Z) = (D,Zg^*)$. Thus we have a semidirect product group

(2.8a) $$G_{s;t,u}(F) = N_{s;t,u}(F) \cdot U(t,u;F).$$

Its product formula is

(2.8b) $$(D_0,Z_0;g_0)(D,Z;g) = (D_0 + D + \mathrm{Im}\ \mathcal{H}(Z_0,Zg_0^*),\ Z_0 + Zg_0^*;\ g_0 g).$$

In particular $G_{s;t,u}(F)$ has center $\mathrm{Im}\ F^{s\times s}$, and the case $s = 1$ gives the semidirect product groups $G_{t,u,F}$ of [30].

The group $GL(s;F)$ of all invertible $s \times s$ matrices over F, also acts by automorphism on $N_{s;t,u}(F)$, by $\gamma(D,Z) = (\gamma D \gamma^*, \gamma Z)$. This commutes with the action of $U(t,u;F)$, so we have a semidirect product

$$(2.9a) \quad P_{s;t,u}(F) = G_{s;t,u}(F) \cdot GL(s;F) = N_{s;t,u}(F) \cdot \{U(t,u;F) \times GL(s;F)\}.$$

Define $GL'(s;F) = \{\gamma \in GL(s;F): \gamma$ preserves Lebesgue measure on $F^s\}$. Denote the multiplication group of positive real numbers by R^+ and view it as the group of positive real scalar matrices in $GL(s;F)$. Then $GL(s;F) = GL'(s;F) \times R^+$, and the maximal unimodular subgroup of $P_{s;t,u}(F)$ is

$$(2.9b) \quad P'_{s;t,u}(F) = N_{s;t,u}(F) \cdot \{U(t,u;F) \times GL'(s;F)\} .$$

The groups (2.9a) give the structure of the maximal parabolics:

2.10 Theorem. *Let* E *be a totally isotropic subspace of dimension* $s > 0$ *in* $F^{p,q}$, *and let* P_E *be the corresponding maximal parabolic subgroup of* $U(p,q;F)$. *There is a Lie group isomorphism*

$$(2.11a) \qquad \varphi: P_{s;p-s,q-s}(F) \cong P_E$$

that carries $N_{s;p-s,q-s}(F)$ *to the unipotent radical and* $U(p-s,q-s;F) \times GL(s;F)$ *to a reductive complement.*

φ *reflects the Langlands decomposition* $P_E = MAN$ *as follows. If* $(p-s,q-s,F) \neq (1,1;R)$, *then*

$$(2.11b) \quad N = \varphi \cdot N_{s;p-s,q-s}(F), \ M = \varphi \cdot \{U(p-s,q-s;F) \times GL'(s;F)\} \ \underline{and} \ A = \varphi \cdot R^+$$

In the exceptional case $(p-s,q-s;F) = (1,1;R)$,

(2.11c) $N = \varphi \cdot N_{s;1,1}(R)$, $M = \varphi \cdot \{(\pm I_2) \times GL'(s;R)\}$ _and_ $A = \varphi \cdot R^+$.

Theorem 2.10 is proved in §3. We close §2 by noting some easy consequences.

2.12. Corollary. _A maximal parabolic subgroup_ P_E _of_ $U(p,q;F)$ _has abelian unipotent radical if, and only if, either_ dim E = 1 _and_ F = R, _or_ p = dim E = q.

{This follows from Theorem 2.10 using (2.6a)}.

2.13. Corollary. _A maximal parabolic subgroup_ P_E of $U(p,q;F)$ _is cuspidal (that is, in_ P_E = MAN, M _has a compact Cartan subgroup) if, and only if,_

  F = R: dim E = 1 _and_ p _or_ q _is odd, or_ dim E = 2 _and_ p _or_ q
         is even

  F = C _or_ Q: dim E = 1.

Proof. $M \cong U(p-s,q-s;F) \times GL'(s;F)$ has a compact Cartan subgroup just when both factors have such a subgroup.

If F = R then $GL'(s;F) = SL^{\pm}(s;R)$ has rank s-1 and its maximal subgroup O(s) has rank $\left[\frac{s}{2}\right]$. Equality holds only for s = 1,2. $U(p-s,q-s;F) = O(p-s,q-s)$ has a compact Cartan subgroup just when (p-s)(q-s) is even.

If $F \neq R$ then $U(p-s,q-s;F)$ always has a compact Cartan subgroup, $GL'(s;F)$ has rank 2s-1 as real Lie group, and the maximal compact subgroup U(s) or Sp(s) in the latter has rank s, so the condition is s = 1.

                                                          _q.e.d._

2.14. Corollary. *Let* E *be a totally isotropic subspace of dimension* s > 0 *in* $F^{p,q}$. *Then the isomorphism of Theorem* 2.10 *restricts to an isomorphism of* $G_{s;p-s,q-s}(F)$ *onto* $\{g \in U(p,q;F): g|_E = identity\}$.

{This will come out of the proof of Theorem 2.10}.

§3. Parabolic Subgroups of Unitary Groups: Proof of Structure

We prove Theorem 2.10 and note the point at which Corollary 2.14 is proved.

Fix an orthonormal basis $\{e_1, \ldots, e_{p+q}\}$ of $F^{p,q}$ such that $\{e_1 + e_{p+1}, \ldots, e_s + e_{p+s}\}$ is a basis of E. Denote

$$(3.1a) \qquad E' = \text{span of } \{e_1 - e_{p+1}, \ldots, e_s - e_{p+s}\},$$

$$(3.1b) \qquad V = E + E' = \text{span of } \{e_1, \ldots, e_s; e_{p+1}, \ldots, e_{p+s}\},$$

$$(3.1c) \qquad W = V^\perp = \text{span of } \{e_{s+1}, \ldots, e_p; e_{p+s+1}, \ldots, e_{p+q}\}.$$

Then $V \cong F^{s,s}$, $W \cong F^{p-s,q-s}$ and $F^{p,q} = V \oplus W$. Denote

$$(3.2a) \qquad \mathfrak{u}: \text{ Lie algebra of } U(p,q;F),$$

$$(3.2b) \qquad \mathfrak{p}: \text{ Lie algebra of } P_E = \{g \in U(p,q;F): g(E) = E\},$$

$$(3.2c) \qquad \mathfrak{l}: \text{ Lie algebra of } L_E = \{g \in U(p,q;F): g|_E = \text{identity}\}.$$

The first step in understanding $\mathfrak{p}$ and $\mathfrak{l}$ is

3.4. *Lemma*. $\mathfrak{p}$ *is real direct sum of its subspaces given by*

$$\mathfrak{p}_r^W = \{\xi \in \mathfrak{u}: \xi V = 0 \underline{\text{ and }} \xi W \subset W\},$$

$$\mathfrak{p}_r^V = \{\xi \in \mathfrak{u}: \xi E \subset E, \xi E' \subset E' \underline{\text{ and }} \xi W = 0\},$$

$$\mathfrak{p}_n^1 = \{\xi \in \mathfrak{u}: \xi E' \subset W, \xi W \subset E \underline{\text{ and }} \xi E = 0\},$$

$$\mathfrak{p}_n^2 = \{\xi \in \mathfrak{u}: \xi E' \subset E, \xi E = 0 \underline{\text{ and }} \xi W = 0\};$$

Joseph A. Wolf

and $\mathfrak{l} = \mathfrak{p}_r^W + \mathfrak{p}_n^1 + \mathfrak{p}_n^2$.

Proof. Let π_V and π_W denote orthogonal projections of $F^{p,q}$ to V and W. If $\xi \in \mathfrak{p}$ then $\xi = (\pi_V + \pi_W)\xi(\pi_V + \pi_W)$ gives us $\xi = \xi_+ + \xi_1$ where

$$\xi_+ = \pi_V\xi\pi_V + \pi_W\xi\pi_W \quad \text{and} \quad \xi_1 = \pi_V\xi\pi_W + \pi_W\xi\pi_V.$$

Now $\xi_+(E) = \pi_V\xi E \subset \pi_V E = E$; and for $v, v' \in V$ and $w, w' \in W$,

$$h(\xi_+(v+w),\ v'+w') + h(v+w,\ \xi_+(v'+w'))$$

$$= h(\xi_+v,v') + h(v,\xi_+v') + h(\xi_+w,w') + h(w,\xi_+w')$$

$$= h(\xi v,v') + h(v,\xi v') + h(\xi w,w') + h(w,\xi w') = 0.$$

Thus $\xi_+ \in \mathfrak{u}$ and $\xi_+(E) \subset E$, that is $\xi_+ \in \mathfrak{p}$, so also $\xi_1 = \xi - \xi_+ \in \mathfrak{p}$.

We check $\mathfrak{p}_n^1 = \{\xi_1 : \xi \in \mathfrak{p}\}$. If $\xi \in \mathfrak{p}_n^1$ then $\xi_1 = \xi \in \mathfrak{p}_n^1$. Now let $\xi \in \mathfrak{p}$. As ξ_1 interchanges W with $V = E + E'$, we need only verify

$$\xi_1 W \subset E \quad \text{and} \quad \xi_1 E = 0.$$

For the second, $\xi_1 E = \pi_W\xi\pi_V E = \pi_V\xi E \subset \pi_W E = 0$. Now

$$0 = h(\xi_1 w, e) + h(w, \xi_1 e) = h(\xi_1 w, e)$$

shows $\xi_1 W \perp E$. As $\xi_1 W \subset V \perp W$ now $\xi_1 W \subset (E+W)^\perp = E$.

Let $\pi_{E'}$ and π_{E+W} denote the projections of $F^{p,q}$ that correspond to the splitting $F^{p,q} = E' + (E+W)$. If $\xi \in \mathfrak{p}$ then $\xi_+ = (\pi_{E'} + \pi_{E+W})\xi_+(\pi_{E'} + \pi_{E+W})$ gives $\xi_+ = \xi_0 + \xi_2$ where

$$\xi_0 = \pi_{E'}\xi_+\pi_{E'} + \pi_{E+W}\xi_+\pi_{E+W} \quad \text{and} \quad \xi_2 = \pi_{E'}\xi_+\pi_{E+W} + \pi_{E+W}\xi_+\pi_{E'}.$$

Notice $\xi_2 E' = \pi_{E+W}\xi_+ E' \subset \pi_{E+W}V = E$ and $\xi_2(E+W) = \pi_{E'}\xi_+(E+W) \subset \pi_{E'}(E+W) = 0$. If $f, f' \in E'$ and $x, x' \in E+W$ now

$$h(\xi_2(f+x), f'+x') + h(f+x, \xi_2(f'+x')) = h(\xi_2 f, f') + h(f, \xi_2 f')$$

$$= h(\pi_{E+W}\xi_+ f, f') = h(f, \pi_{E+W}\xi_+ f') = h(\xi_+ f, f') + h(f, \xi_+ f') = 0.$$

We have just proved $\xi_2 \in \mathfrak{p}_n^2$, and it follows that $\mathfrak{p}_n^2 = \{\xi_2 : \xi \in \mathfrak{p}\}$.

Let $\mathfrak{p}_r = \mathfrak{p}_r^V + \mathfrak{p}_r^W$. We check $\mathfrak{p}_r = \{\xi_0 : \xi \in \mathfrak{p}\}$. If $\xi \in \mathfrak{p}$, then $\xi_0 = \xi_+ - \xi_2 \in \mathfrak{p}$. $\xi_0 E' \subset E'$ by construction. Now ξ_0 preserves $V = E + E'$, thus also $W = V^\perp$, so $\xi_0 \in \mathfrak{p}_r$. Conversely, if $\xi \in \mathfrak{p}_r$ it preserves E, E' and W, hence also V and $E + W$, so $\xi_0 = \xi \in \mathfrak{p}_r$.

Now $\mathfrak{p} = \mathfrak{p}_r + \mathfrak{p}_n^1 + \mathfrak{p}_n^2$, vector space sum with the splitting $\xi = \xi_0 + \xi_1 + \xi_2$. As $\mathfrak{p}_r = \mathfrak{p}_r^V + \mathfrak{p}_r^W$ direct sum of subspaces, we have decomposed $\mathfrak{p}$ as required.

Evidently, $\mathfrak{p}_r^W + \mathfrak{p}_n^1 + \mathfrak{p}_n^2 \subset \mathfrak{l}$. If $\xi \in \mathfrak{l} \cap \mathfrak{p}_r^V$ then for $e \in E$ and $e' \in E'$, $0 = h(\xi e, e') + h(e, \xi e') = h(e, \xi e')$. As $\xi e' \in E'$ which pairs nonsingularly with E, now $\xi e' = 0$. Thus $\xi E' = 0$ and so $\xi = 0$. Thus $\mathfrak{l} \cap \mathfrak{p}_r^V = 0$ and $\mathfrak{l}$ is decomposed as required.

<u>q.e.d.</u>

If we apply (3.3a) to the Lie algebra $\mathfrak{u}(p-s, q-s; F)$ of the unitary group $U(p-s, q-s; F)$ of $W \cong F^{p-s, q-s}$, we see

$$(3.5a) \qquad \mathfrak{u}(p-s, q-s; F) = \left\{ \begin{pmatrix} \alpha & \beta \\ \beta^* & \delta \end{pmatrix} : \alpha + \alpha^* = 0 \text{ and } \delta + \delta^* = 0 \right\}$$

where $\alpha \in F^{(p-s) \times (p-s)}$, $\beta \in F^{(p-s) \times (q-s)}$ and $\delta \in F^{(q-s) \times (q-s)}$. We also represent the Lie algebra of the group $GL(s; F)$ of all invertible $s \times s$

matrices over F, by

(3.5b) $\mathfrak{gl}(s;F) = \{\gamma: \gamma \in F^{s \times s}\}$ as usual.

Now we can identify $\mathfrak{p}_r$:

      3.6. Lemma. *Define* ψ_r: $\mathfrak{u}(p-s,q-s;F) \oplus \mathfrak{gl}(s;F) \to \mathfrak{gl}(p+q;F)$ *by*

$$\psi_r\left(\begin{pmatrix} \alpha & \beta \\ \beta* & \delta \end{pmatrix}, \gamma\right) = \begin{bmatrix} \text{Im } \gamma & 0 & \text{Re } \gamma & 0 \\ 0 & \alpha & 0 & \beta \\ \text{Re } \gamma & 0 & \text{Im } \gamma & 0 \\ 0 & \beta* & 0 & \delta \end{bmatrix} \;.$$

Then $\psi_r \cdot \mathfrak{u}(p-s,q-s;F) = \mathfrak{p}_r^W$, $\psi_r \cdot \mathfrak{gl}(s;F) = \mathfrak{p}_r^V$, *and* ψ_r *is a Lie algebra isomorphism onto* $\mathfrak{p}_r = \mathfrak{p}_r^W \oplus \mathfrak{p}_r^V$.

    Proof.   Let $\xi \in \mathfrak{p}_r$. Then $\xi W \subset W$ with $\xi|_W$ in the Lie algebra of the unitary group there, say $\xi|_W$ matrix $\begin{pmatrix} \alpha & \beta \\ \beta* & \delta \end{pmatrix} \in \mathfrak{u}(p-s,q-s;F)$. Similarly, $\xi|_V$ has matrix $\begin{pmatrix} \rho & \sigma \\ \sigma* & \tau \end{pmatrix} \in \mathfrak{u}(s,s;F)$. $\xi E \subset E$ just says $\rho + \sigma = \sigma* + \tau$, and $\xi E' \subset E'$ is equivalent to $\rho - \sigma = \tau - \sigma*$. Together these say $\rho = \tau$, $\sigma = \sigma*$ and $\rho + \rho* = 0$.   Thus

$$\xi|_V \quad \text{has matrix} \quad \begin{bmatrix} \text{Im } \gamma & \text{Re } \gamma \\ \text{Re } \gamma & \text{Im } \gamma \end{bmatrix} \quad \text{where} \quad \gamma = \sigma + \rho.$$

Now it is clear that ψ_r is a vector space isomorphism of $\mathfrak{u}(p-s,q-s;F) \oplus \mathfrak{gl}(s;F)$ onto $\mathfrak{p}_r$ which restricts to a Lie algebra isomorphism from $\mathfrak{u}(p-s,q-s;F)$ to $\mathfrak{p}_r^W$ and sends $\mathfrak{gl}(s;F)$ to $\mathfrak{p}_r^V$. To see that the latter is a Lie algebra homomorphism, note that

$$[\gamma,\gamma'] = (\text{Re } \gamma + \text{Im } \gamma)(\text{Re } \gamma' + \text{Im } \gamma') - (\text{Re } \gamma' + \text{Im } \gamma')(\text{Re } \gamma + \text{Im } \gamma)$$

has hermitian part $(\text{Im } \gamma)(\text{Re } \gamma') + (\text{Re } \gamma)(\text{Im } \gamma') - (\text{Im } \gamma')(\text{Re } \gamma) -$
$(\text{Re } \gamma')(\text{Im } \gamma)$ and skew hermitian part $(\text{Im } \gamma)(\text{Im } \gamma') + (\text{Re } \gamma)(\text{Re } \gamma') -$
$(\text{Re } \gamma')(\text{Re } \gamma) - (\text{Im } \gamma')(\text{Im } \gamma)$, so that $\gamma \to \begin{pmatrix} \text{Im } \gamma & \text{Re } \gamma \\ \text{Re } \gamma & \text{Im } \gamma \end{pmatrix}$ sends bracket to bracket.

$$q.e.d.$$

$\underline{3.7.}$ $\underline{\text{Lemma.}}$ $\textit{Define}$ $\psi_n\colon \mathfrak{n}_{s;p-s,q-s}(F) \to \mathfrak{gl}(p+q;F)$ $\textit{by}$

$$\psi_n(D,(A,B)) = \begin{bmatrix} -D & A & D & B \\ -A^* & 0 & A^* & 0 \\ -D & A & D & B \\ B^* & 0 & -B^* & 0 \end{bmatrix} \ .$$

$\underline{\textit{Then}}$ $\psi_n \cdot \text{Im } F^{s \times s} = \mathfrak{p}_n^2$, $\psi_n \cdot F^{s \times (p-s,q-s)} = \mathfrak{p}_n^1$ $\underline{\textit{and}}$ ψ_n $\underline{\textit{is a Lie algebra}}$
$\underline{\textit{isomorphism}}$ $\underline{\textit{onto}}$ $\mathfrak{p}_n = \mathfrak{p}_n^2 + \mathfrak{p}_n^1$. $\underline{\textit{Further, the image of}}$ ψ_n $\underline{\textit{consists of}}$
$\underline{\textit{nilpotent matrices}}$.

$\underline{\textit{Proof}}$. Denote $\zeta_D = \psi_n(D,0)$ and $\eta_{A,B} = \psi_n(0,(A,B))$. By direct calculation,

$$\zeta_{D_0}\zeta_D = 0, \quad \text{in particular} \quad \zeta_D^2 = 0 \quad \text{and} \quad [\zeta_{D_0},\zeta_D] = 0.$$

Again, by direct calculation,

$$\zeta_D\eta_{A,B} = 0 = \eta_{A,B}\zeta_D, \quad \text{in particular} \quad [\zeta_D,\eta_{A,B}] = 0.$$

Finally, by direct calculation with $Z_0 = (A_0,B_0)$ and $Z = (A,B)$,

$$\eta_{A_0,B_0}\,\eta_{A,B} = \begin{bmatrix} -\mathcal{H}(Z_0,Z) & 0 & \mathcal{H}(Z_0,Z) & 0 \\ 0 & 0 & 0 & 0 \\ -\mathcal{H}(Z_0,Z) & 0 & \mathcal{H}(Z_0,Z) & 0 \\ 0 & 0 & 0 & 0 \end{bmatrix} .$$

In particular,

$$\eta_{A,B}^3 = 0 \quad \text{and} \quad [\eta_{A_0,B_0}, \eta_{A,B}] = \zeta_{2\text{Im } \mathcal{H}(Z_0,Z)}.$$

We conclude that ψ_n is a Lie algebra isomorphism onto its image and the image consists of nilpotent matrices. Now we must identify that image.

Let $\zeta \in \mathfrak{p}_n^2$. As $\zeta E' \subset E$ and $\zeta(E + W) = 0$, ζ has matrix of the form

$$\begin{bmatrix} -D & 0 & D & 0 \\ 0 & 0 & 0 & 0 \\ -D & 0 & D & 0 \\ 0 & 0 & 0 & 0 \end{bmatrix}$$

and $\zeta \in \mathfrak{u}$ just says $D + D^* = 0$. Conversely, any such matrix is in $\mathfrak{p}_n^2$. We have shown $\psi_n \cdot \text{Im } F^{s \times s} = \mathfrak{p}_n^2$.

Let $\eta \in \mathfrak{p}_n^1$. As η interchanges $V = E + E'$ and W, and $\eta E = 0$, η has matrix of the form

$$\begin{bmatrix} 0 & A & 0 & B \\ -C & 0 & C & 0 \\ 0 & A' & 0 & B' \\ -C' & 0 & C' & 0 \end{bmatrix}.$$

But $\eta \in \mathfrak{u}$, so $C = A^*$, $A' = C^* = A$, $C' = -B^*$ and $B' = -C'^* = B$. Thus $\eta = \eta_{A,B}$ in the image of ψ_n. Conversely any such matrix is in $\mathfrak{p}_n^1$. We have shown $\psi_n \cdot F^{s \times (p-s, q-s)} = \mathfrak{p}_n^1$.

$$\underline{q.e.d.}$$

Lemmas 3.6 and 3.7 give us a real vector space isomorphism

$$\psi \colon \mathfrak{p}_{s;p-s,q-s}(F) \to \mathfrak{p} \quad \text{such that} \quad \psi \cdot \mathfrak{g}_{s;p-s,q-s}(F) = \mathfrak{l} \quad \text{and the restrictions}$$

$$\psi_n \colon \mathfrak{n}_{s;p-s,q-s}(F) \to \mathfrak{p}_n \quad \text{and} \quad \psi_r \colon \mathfrak{u}(p-s,q-s;F) \oplus \mathfrak{gl}(s;F) \to \mathfrak{p}_r$$

are Lie algebra isomorphisms. Now we must check that ψ preserves Lie algebra products between $\mathfrak{n}_{s;p-s,q-s}(F)$ and $\mathfrak{u}(p-s,q-s;F) \oplus \mathfrak{gl}(s;F)$. Direct calculation gives

3.8. Lemma. *Let* $\begin{pmatrix} \alpha & \beta \\ \beta* & \delta \end{pmatrix} \in u(p-s,q-s;F)$, $\gamma \in gl(s;F)$, $D \in \text{Im } F^{s \times s}$ *and* $\zeta_D = \psi_n(D,0)$, *and* $Z = (A,B) \in F^{s \times (p-s,q-s)}$ *and* $\eta_Z = \eta_{A,B} = \psi_n(0,(A,B))$ *as in Lemma 3.7. Then*

(3.9a) $[\psi_r(\begin{pmatrix} \alpha & \beta \\ \beta* & \delta \end{pmatrix},0), \zeta_D] = 0$ *and* $[\psi_r(\begin{pmatrix} \alpha & \beta \\ \beta* & \delta \end{pmatrix},0), \eta_{A,B}] = \eta_{A',B'}$

with $(A',B') = -(A,B)\begin{pmatrix} \alpha & \beta \\ \beta* & \delta \end{pmatrix}$, *and*

(3.9b) $[\psi_r(0,\gamma),\zeta_D] = \zeta_{\gamma D + D \gamma*}$ *and* $[\psi_r(0,\gamma),\eta_Z] = \eta_{\gamma Z}$.

Comparing Lemma 3.8 with the discussions just preceding the definitions (2.8) and (2.9), we conclude that ψ preserves all Lie algebra products:

3.10. Proposition. *Let* $\psi = \psi_n \oplus \psi_r$ *in the notation of Lemma 3.6 and 3.7. Then* $\psi: p_{s;p-s,q-s}(F) \to p$ *is a Lie algebra isomorphism which restricts to an isomorphism* $g_{s;p-s,q-s}(F)$ *onto* l.

Theorem 2.10 and Corollary 2.14 now are proved on the Lie algebra level.

P_E is a real linear algebraic subgroup of $U(p,q;F)$ by its definition. Lemmas 3.6, 3.7 and 3.8 show that p_n is the Lie algebra of its unipotent radical and p_r is the Lie algebra of an R-rational reductive complement; for p_n and p_r are defined by where they send certain subspaces of $F^{p,q}$, p_n is an ideal consisting of nilpotent transformations, and p_r is a completely reducible complementary subalgebra of p. Both $N_{s;p-s,q-s}(F)$ and the nilpotent radical of P_E are connected simply connected Lie groups. So ψ_n exponentiates:

3.11. Lemma. *The isomorphism* ψ_n *of* $n_{s;p-s,q-s}(F)$ *onto* p_n *is induced by an isomorphism* $\varphi_n: N_{s;p-s,q-s}(F) \cong$ *(unipotent radical of* P_E*).*

ψ_r is a bit more difficult of exponentiate because we may have neither connectivity nor simple connectivity for our reductive groups. Let us start with the simple fact

(3.12)
$$\begin{cases} \underline{Let} \quad H = \{g: V \to V \quad \underline{unitary:} \ gE = E \ \underline{and} \ gE' = E'\} \quad . \quad \underline{Then}^* \\ g \to g|_E \quad \underline{maps} \ H \quad \underline{isomorphically \ to \ the \ general \ linear \ group \ of} \ E. \end{cases}$$

This allows us to state

    3.13. Lemma. *Define* φ_r: $U(p-s,q-s;F) \times GL(s;F) \to U(p,q;F)$ *by*

(i) $\varphi_r|_{U(p-s,q-s;F)}$ *is the isomorphism of* $U(p-s,q-s;F)$ *onto the unitary group* $\{g \in U(p,q;F): g|_V = 1 \ \underline{and} \ gW = W\}$ *of* W *that is induced by the isometry* $x \to \sum_{\ell=1}^{p-s} e_{s+\ell} x^\ell + \sum_{m=1}^{q-s} e_{p+s+m} x^{p-s+m}$ *of* $F^{p-s,q-s}$ *onto* W.

(ii) $\varphi_r|_{GL(s;F)}$ *is the isomorphism of* $GL(s;F)$ *onto* $\{g \in U(p,q;F): g|_W = 1, \ gE = E \ \underline{and} \ gE' = E'\}$ *specified by* (3.12) *and the linear isomorphism* $y \to \sum_{m=1}^{s}(e_m + e_{p+m})y^m$ *of* F^s *onto* E.

Then φ_r *induces the Lie algebra isomorphism* ψ_r *of Lemma 3.6, and* φ_r *is an isomorphism of* $U(p-s,q-s;F) \times GL(s;F)$ *onto* $\{g \in P_E: g\mathfrak{p}_r g^{-1} = \mathfrak{p}_r\}$.

    Proof. φ_r induces ψ_r by construction, φ_r is an isomorphism onto image(φ_r), and image(φ_r) $\subset P_E$. It follows that image(φ_r) is contained

---

* Restriction to E is a continuous homomorphism from H, so one need only see that it is bijective. If $g|_E = 1$ then $g|_{E'} = 1$ because $h(e,ge') = h(ge,ge') = h(e,e')$ for $e \in E$ and $e' \in E'$ and because $h: E \times E' \to F$ is a nonsingular hermitian pairing. Thus $g \to g|_E$ is injective. If $k \in GL(E)$ has matrix γ in the basis $\{e_i + e_{p+i}\}$. Define $k' \in GL(E')$ to have matrix $(\gamma^*)^{-1}$ in the basis $\{e_i - e_{p+i}\}$, and then $g(e + e') = k(e) + k'(e')$ defines $g \in H$ with $g|_E = k$. Thus $g \to g|_E$ is surjective.

in the P_E-normalizer of image(ψ_r) = $\mathfrak{p}_r$. Now we need only verify that the

P_E-normalizer of $\mathfrak{p}_r$ is contained in image(φ_r).

Let $g \in P_E$ with $g\mathfrak{p}_r g^{-1} = \mathfrak{p}_r$. We are going to see that $gV = V$ and

$gW = W$. Let α be the set of all central elements $\xi \in \mathfrak{p}_r$ such that ξ

has real diagonal matrix in some basis of $F^{p,q}$. Then $g\alpha g^{-1} = \alpha$, and

$\alpha = \alpha_V + \alpha_W$ where

$$\alpha_V = \alpha \cap \psi_r \cdot \mathfrak{gl}(s;F) \quad \text{and} \quad \alpha_W = \alpha \cap \psi_r \cdot \mathfrak{u}(p-s,q-s;F).$$

Since $s \geqslant 1$, $\alpha_V = \psi_r$ (real scalar $s \times s$ matrices) consists of all real

multiples of a certain nonzero element $\xi_V = \psi_r(I_s)$, whose null space is W.

If $(p-s,q-s;F) \neq (1,1;R)$, then $\alpha_W = 0$, so $g\xi_V g^{-1}$ is a nonzero multiple

of ξ_V, forcing $gW = W$ and thus also $gV = V$. In the exceptional case

$(p-s,q-s;F) = (1,1;R)$, $\alpha_W = \psi_r \cdot \mathfrak{u}(1,1;R)$, which consists of all real multi-

ples of $\xi_W = \psi_r\begin{pmatrix}0 & 1\\1 & 0\end{pmatrix}$, whose null space is V. Now α consists of the

$a\xi_V + b\xi_W$, a and b real, and $a\xi_V + b\xi_W$ has nonzero null space only

when ab = 0. If $g\xi_V g^{-1}$ is not a multiple of ξ_V, now it is a mutiple

of ξ_W, so g interchanges V and W, contradicting $g \in P_E$. Now

$gW = W$ and $gV = V$ as before.

Let $g \in P_E$ with $g\mathfrak{p}_r g^{-1} = \mathfrak{p}_r$. As $gV = V$ and $gW = W$ we have

$g = g_V g_W$ where

$$g_V(v+w) = g(v) + w \quad \text{and} \quad g_W(v+w) = v + g(w)$$

for $v \in V$ and $w \in W$. It is immediate that $g \in$ image(φ_r).

q.e.d.

Now $P = P_E$ has semidirect product decomposition $P = P_n \cdot P_r$ where the

unipotent radical $P_n =$ image(φ_n) and the Chevalley reductive complement

$P_r =$ image(φ_r). A glance back at the construction (2.8a), (2.9a) of

$P_{s;p-s,q-s}(F)$ and at the matrices in Lemmas 3.6, 3.7, 3.8 and 3.13 now gives

Joseph A. Wolf

us the isomorphism $\varphi: P_{s;p-s,q-s}(F) \to P_E$ by

$\varphi((D,(A,B)),g_W,\gamma) = \varphi_n((D,(A,B)) \cdot \varphi_r(g_W,\gamma)$. With the discussion of the

structure of α in the second paragraph of the proof of Lemma 3.13, this

completes the proof of Theorem 2.10 and Corollary 2.14.

§4:   Unitary Representations of the Nilradical

We now work out the irreducible unitary representations of the groups $N_{s;t,u}(F)$. This is a straightforward application of the Kirillov theory([4]; or see [18]).

It will be convenient to use the notation

(4.1a) $N = N_{s;t,u}(F)$,   our simply connected nilpotent group;

(4.1b) $\mathfrak{n} = \mathfrak{n}_{s;t,u}(F) = \text{Im } F^{s\times s} + F^{s\times(t,u)}$,   its Lie algebra;

(4.1c) $\mathfrak{n}^*$:   the real linear dual space of $\mathfrak{n}$;   and

(4.1d) Ad^*:   the (co-adjoint) representation of N on $\mathfrak{n}^*$.

The set $\hat{N}$ of equivalence classes of irreducible unitary representations of N is in bijective correspondence with the set of all $\text{Ad}^*(N)$-orbits on $\mathfrak{n}^*$, so we want to investigate the orbit structure on $\mathfrak{n}^*$.

If M is a square matrix (m^i_j),   then trace M means the sum $\sum m^i_i$ of the diagonal elements. Note trace Re $M = $ Re trace M. Now define a real pairing on $\mathfrak{n}$ by the formula

(4.2a) $<(D_0,Z_0),(D,Z)> = \text{trace Re } D_0 D^* + \text{trace Re } \mathcal{H}(Z_0,Z)$.

This is easily seen nondegenerate on $\text{Im } F^{s\times s}$ and $F^{s\times(t,u)}$ separately, and $<\text{Im } F^{s\times s},\ F^{s\times(t,u)}> = 0$,   so (4.2a)   is a nondegenerate symmetric real bilinear form on $\mathfrak{n}$. Thus

(4.2b) $\mathfrak{n}^* = \{f_{D,Z}:\ (D,Z) \in \mathfrak{n}\}$ where $f_{D,Z}(D_0,Z_0) = <(D_0,Z_0),\ (D,Z)>$.

It will sometimes be convenient to write f_D for $f_{D,0}$ and g_Z for $f_{0,Z}$, and we will do this without further comment.

    4.3. Proposition. *If* $D \in \text{Im } F^{s \times s}$ *then* $F^{s \times (t,u)}$ *splits as direct sum of real subspaces orthogonal for the inner product* (4.2a),

$$(4.4a) \qquad F^{s \times (t,u)} = \{Z \in F^{s \times (t,u)} : DZ = 0\} \oplus \{DZ : Z \in F^{s \times (t,u)}\}.$$

If $f = f_{D,Z} \in \mathfrak{n}^*$, *then the co-adjoint orbit is the affine subspace*

$$(4.4b) \qquad \text{Ad}^*(N) \cdot f = f + \{g_{DZ'} : Z' \in F^{s \times (t,u)}\}$$

of $\mathfrak{n}^*$, *and the isotropy algebra* $\mathfrak{n}_f = \{x \in \mathfrak{n} : f[x,\mathfrak{n}] = 0\}$ *is given by*

$$(4.4c) \qquad \mathfrak{n}_f = \text{Im } F^{s \times s} + \{Z' \in F^{s \times (t,u)} : DZ' = 0\}.$$

    Proof. Every $D \in \text{Im } F^{s \times s}$ being skew-hermitian, the linear transformation $Z \to DZ$ of $F^{s \times (t,u)}$ is completely reducible, so $F^{s \times (t,u)}$ is direct sum (4.4a) of kernel and image. If Z, $Z' \in F^{s \times (t,u)}$ with $DZ = 0$, then the inner product

$$<Z, DZ'> = \text{trace Re } \mathcal{H}(Z, DZ') = -\text{trace Re } \mathcal{H}(DZ, Z') = 0$$

so the summands are othogonal.

    Let $f = f_{D,Z} \in \mathfrak{n}^*$ and compute $f[(D',Z'), (D'',Z'')]$ $= f(2 \text{ Im } \mathcal{H}(Z',Z''),0) = \text{trace Re}\{2 \text{ Im } \mathcal{H}(Z',Z'')D^*\}$ as follows. If $Z'' = (A'',B'')$, let $Z_1 = (A'',-B'')$, so that $\mathcal{H}(Z',Z'') = Z'Z_1^*$. Then

$$2 \text{ Im } \mathcal{H}(Z',Z'')D^* = (Z'Z_1^* - Z_1Z'^*)D^*$$

and the real component of its trace is

$$\text{Re trace}(Z'Z_1^*D^*) - \text{Re trace}(Z_1Z'^*D^*)$$

$$= \text{Re trace}(D^*Z'Z_1^*) - \text{Re trace}(DZ'Z_1^*)^*$$

$$= -\{\text{Re trace}(DZ'Z_1^*) + \text{Re trace}(DZ'Z_1^*)^*\}$$

$$= -2\,\text{trace Re}(DZ'Z_1^*) = -2\langle DZ', Z'\rangle.$$

In view of (4.4a), now $(D',Z') \in n_f$ if and only if $DZ' = 0$. This proves (4.4c).

The orbit $\text{Ad}^*(N) \cdot f$ has real tangent space at f given by

$$T_f(\text{Ad}^*(N)\cdot f) = f + \{f'' \in n^* : f''(n_f) = 0\}$$

$$= f + \{f_{D'',Z''} \in n^* : \langle D'',Z''), n_f\rangle = 0\}$$

$$= f + \{g_{Z''} \in n^* : Z'' \perp \{Z':DZ' = 0\}\} = f + \{g_{DZ'} : Z' \in F^{s\times(t,u)}\},$$

using (4.4c) and then (4.4a). But the orbit is an affine subspace of n^* because N is 2-step nilpotent or abelian. Thus the orbit is equal to its tangent space and (4.4b) is proved.

$$q.e.d.$$

4.5. _Corollary_. $N_{s;t,u}(F)$ _has square integrable representations_ (_see_ [17]) _except in the case where_ $F = R$ _and_ s _is odd with_ $s \geqslant 3$.

Proof. The representation class $[\pi_f]$ for the orbit of $f = f_{D,Z} \in n^*$ is square integrable if and only if n_f is reduced to the center of n. This is automatic if N is abelian (2.6); in the non-abelian case it is equivalent to invertibility of D.

$$q.e.d.$$

Given $f = f_{D,Z} \in \mathfrak{n}^*$ we denote

(4.6a) $\mathfrak{d}_f = \{D' \in \mathrm{Im}\ F^{s \times s} : f(D') = 0\}$,   central ideal in $\mathfrak{n}$;

(4.6b) $\mathfrak{m}_f = \mathfrak{n}/\mathfrak{d}_f$ quotient algebra and $p: \mathfrak{n} \to \mathfrak{m}_f$ projection;

(4.6c) $\mathfrak{h}_f = p(\mathrm{Im}\ F^{s \times s} + D \cdot F^{s \times (t,u)})$;   and

(4.6d) $\alpha_f = p(\mathfrak{d}_f + \{Z' \in F^{s \times (t,u)} : DZ' = 0\})$.

Then $\mathfrak{h}_f$ and α_f are subalgebras of $\mathfrak{m}_f$, and we have

(4.7a) M_f: the simply connected nilpotent Lie group for $\mathfrak{m}_f$; and

(4.7b) H_f and A_f: the analytic subgroups of M_f for $\mathfrak{h}_f$ and α_f.

It is clear that A_f is abelian, hence is a real vector group. The possibilities for H_f are

(4.8a) H_f is a Heisenberg group of dimension $\geqslant 3$ ($D \neq 0$, $t + u > 0$),

(4.8b) H_f is the 1-dimensional real vector group ($D \neq 0$, $t = u = 0$),

(4.8c) H_f is the trivial group ($D = 0$).

In any case, we note $\mathfrak{m}_f = \mathfrak{h}_f + \alpha_f$, $\mathfrak{h}_f \cap \alpha_f = 0$ and $[\mathfrak{h}_f, \alpha_f] = 0$,   so

(4.9) $M_f = H_f \times A_f$.

The Kirillov theory tells us that the unitary representation class $[\pi_f] \in \hat{N}$ for the orbit of f is given by

(4.10) $[\pi_f] = [\bar{\pi}_f \cdot p]$ where $[\bar{\pi}_f] \in \hat{M}_f$ is the class for $\bar{f}$ with $f = \bar{f} \cdot p$, because the kernel $\exp(\delta_f)$ of $p: N \to M_f$ is a central subgroup contained in the kernel of π_f. Since our groups are type I, (4.9) says

(4.11) $[\bar\pi_f] = [\eta_f \otimes \alpha_f]$ where $[\eta_f] \in \hat{H}_f$ and $\alpha_f \in \hat{A}_f$.

Evidently, α_f is the unitary character

(4.12) $\alpha_f = e^{ig_Z}: \exp(\delta_f + Z')_- \to e^{ig_Z(Z')} = e^{i \text{ trace Re } \mathcal{H}(Z',Z)}.$

Further, η_f acts on $p(\exp \text{ Im } F^{s \times s})$ by the unitary character

$$e^{if_D}: \exp(\delta_f + D') \to e^{if_D(D')} = e^{i \text{ trace Re } D'D^*}.$$

Now, according to the three cases of (4.8),

(4.13) H_f is a Heisenberg group and $[\eta_f] \in \hat{H}_f$ has central character $e^{if_D} \neq 1$ ($D \neq 0$, $t + u > 0$),

(4.13b) $H_f \cong R^1$ and η_f is its character $e^{if_D} \neq 1$ ($D \neq 0$, $t + u = 0$),

(4.13c) H_f and η_f are trivial ($D = 0$).

We summarize and complete the discussion to this point:

4.14. Theorem. *Let* $N = N_{s;t,u}(F)$. *Then* $\hat{N}$ *consists of the classes* $[\pi_{D,Z}] = [\pi_{f_{D,Z}}]$ *given as follows.*

(4.15a) $D \in \text{Im } F^{s \times s}$ *and* $Z \in F^{s \times (t,u)}$ *with* $DZ = 0$,

(4.15b) $p: N \to M_f = H_f \times A_f$ _as in_ (4.6) - (4.9) _with_ $f = f_{D,Z}$,

(4.15c) $\alpha_f = e^{ig_Z} \in \hat{A}_f$ _is the unitary character_ (4.12),

(4.15d) $[\eta_f] \in \hat{H}_f$ _is the class given by_ (4.8) _and_ (4.13),

(4.15e) $[\pi_{D,Z}] = [(\eta_f \otimes \alpha_f) \cdot p] \in \hat{N}$.

In addition,

   (i) $[\pi_{D,Z}] = [\pi_{D',Z'}]$ _if and only if_ $(D,Z) = (D',Z')$,

   (ii) _the central character of_ $[\pi_{D,Z}]$ _restricts to_ e^{if_D} _on_ $\text{Im } F^{s \times s}$,

   (iii) _if_ $D \neq 0$ _then_ $[\pi_{D,Z}]$ _is infinite dimensional, and_

   (iv) _if_ $D = 0$ _then_ $[\pi_{D,Z}]$ _is a unitary character._

Finally, if N _is nonabelian, then the Plancherel measure on_ $\hat{N}$ _is concentrated on_ $\{[\pi_{D,Z}]: D$ _has rank_ s' _as a matrix_$\}$ _where_ s' _is maximal:_ $s' = s-1$ _if_ $F = R$ _and_ s _is odd,_ $s' = s$ _otherwise._

   Proof. The Kirillov Theory ensures that $\hat{N}$ consists of the representation classes $[\pi_{f_{D,Z}}]$ associated to the co-adjoint orbits $\text{Ad}^*(N) \cdot f_{D,Z}$, and our discussion shows that $[\pi_{f_{D,Z}}]$ is given by (4.15b,c,d,e). Furthermore, using Proposition 4.3,

$$[\pi_{f_{D,Z}}] = [\pi_{f_{D',Z'}}] \Longleftrightarrow (D',Z') \in \text{Ad}^*(N) \cdot f_{D,Z} \Longleftrightarrow D = D' \text{ and } Z-Z' \in D \cdot F^{s \times (t,u)}$$

Thus $[\pi_{f_{D,Z}}]$ has unique such expression if we add the condition (4.15a) that $DZ = 0$. This also proves (i); and (ii), (iii) and (iv) follow from (4.13).

The Plancherel measure on $\hat{N}$ is concentrated on $\{[\pi_f]: \mathrm{Ad}^*(N) \cdot f$ has maximal dimension$\}$. This says nothing in the abelian case, but in the non-abelian case it says that D has maximal possible rank (as a matrix). That rank is s', as specified.

q.e.d.

§5. Representations of the groups $G_{s;t,u}(F)$.

We use the results of §4 and the Mackey little group method to work out
the irreducible unitary representations of the groups
$G_{s;t,u}(F) = N_{s;t,u}(F) \cdot U(t,u;F)$. Here the main points are the computation of
the stabilizers

$$(5.1a) \qquad L_{D,Z} = \{g \in U(t,u;F): \ \pi_{D,Z} \cdot Ad(g^{-1}) \in [\pi_{D,Z}]\},$$

which gives us the Mackey little groups

$$(5.1b) \qquad N_{s;t,u}(F) \cdot L_{D,Z}, \quad \text{the} \ G_{s;t,u}(F)\text{-normalizer of} \ [\pi_{D,Z}],$$

and the extension of $[\pi_{D,Z}]$ to a class $[\tilde{\pi}_{D,Z}] \in (N_{s;t,u}(F) \cdot L_{D,Z})^{\wedge}$. The
stabilizers $L_{D,Z}$ are obtained using the structural results of §§2 and 3,
and the extensions $[\tilde{\pi}_{D,Z}]$ are obtained by a geometric realization procedure
that avoids the problem of calculating the Mackey obstruction.

5.2. Lemma. *The co-adjoint action of* $U(t,u;F) \times GL(s;F)$ *on the real
linear dual space of the Lie algebra of* $N_{s;t,u}(F)$, *is given by*
$(g,\gamma)^{-1} \cdot f_{D,Z} = f_{\gamma^*D\gamma, \gamma^*Zg}$.

Proof. If α is an automorphism of N then α acts on $\mathfrak{n}^*$ by
$\alpha(f)(D_0,Z_0) = f(\alpha^{-1} \cdot (D_0,Z_0))$. The case where α is conjugation by
$(g,\gamma)^{-1} \in U(t,u;F) \times GL(s;F)$ gives

$$\{(g,\gamma)^{-1} \cdot f_{D,Z}\}(D_0,Z_0) = f_{D,Z}(\gamma D_0 \gamma^*, \gamma Z_0 g^*)$$

$$= Re \ trace(\gamma D_0 \gamma^* D^*) + Re \ trace \ \mathcal{H}(\gamma Z_0 g^*, Z)$$

$$= Re \ trace(D_0 \cdot \gamma^* D^* \gamma) + Re \ trace \ \mathcal{H}(Z_0, \gamma^* Zg)$$

$$= f_{\gamma^*D\gamma, \gamma^*Zg}(D_0,Z_0).$$

q.e.d.

Let $N = N_{s;t,u}(F)$ and $U = U(t,u;F)$. Glance back at Proposition 4.3 to see that the U-stabilizer of $Ad^*(N) \cdot f_{D,Z}$ is

(5.3a) $L_{D,Z} = \{g \in U : Z - Zg \in D \cdot F^{s \times (t,u)}\}$.

In view of (4.4a), the normalization (4.15a) of Theorem 4.14 simplifies that to

(5.3b)   if $DZ = 0$ then $L_{D,Z} = \{g \in U : Zg = Z\}$,

because $Z - Zg$ is in the intersection of the direct summands (4.4a) of $F^{s \times (t,u)}$. Expressing this in terms of the column span of Z^* and using the fact that $L_{D,Z}$ is the U-stabilizer of $[\pi_{f_{D,Z}}]$, we formulate this as

5.4. Lemma. *Let* $D \in Im\ F^{s \times s}$ *and* $Z \in F^{s \times (t,u)}$ *with* $DZ = 0$. *Denote*

(5.5a) S_Z: *linear subspace of* $F^{t,u}$ *spanned by the columns of* Z^*.

Then the $U(t,u;F)$-*stabilizer of* $[\pi_{D,Z}] \in N_{s;t,u}(F)^{\wedge}$ *is*

(5.5b) $L_{D,Z} = \{g \in U(t,u;F) : g(x) = x$ *for all* $x \in S_Z\}$.

This gives us the algebraic structure of $L_{D,Z}$ by a small variation on Corollary 2.14:

5.6. Proposition. *Let* S *be a linear subspace* $F^{t,u}$, $c = dim(S \cap S^{\perp})$, *and* $a, b \geqslant 0$ *the integers such that* $S \cong (S \cap S^{\perp}) \oplus F^{a,b}$.   *Then*

$$\{g \in U(t,u;F): g(x) = x, \quad \text{all} \quad x \in S\} \cong G_{c;t-c-a,u-c-b}(F).$$

Proof. Let $E = S \cap S^\perp$ and let T be a complement to E in S. Then $S = E \oplus T$ and $T \cap T^\perp = 0$, so $F^{t,u} = T \oplus T^\perp$ with $E \subset T^\perp$. By nondegeneracy, $T \cong F^{a,b}$, so we may identify $T^\perp$ with $F^{t-a,u-b}$. Now $\{g \in U(t,u;F): g|_S = \text{identity}\}$ is identified with $\{g \in U(t-a,t-b;F): g|_E = \text{identity}\}$, where E is a totally isotropic subspace of dimension c in $F^{t-a,u-b}$. Thus our assertion is just Corollary 2.14 with $s = c$, $p = t-a$ and $q = u-b$.

$$q.e.d.$$

Now we know the Mackey little group $N_{s;t,u}(F) \cdot L_{D,Z}$ for $[\pi_{D,Z}]$, and we proceed to extend $[\pi_{D,Z}]$ from $N_{s;t,u}(F)$ to the little group.

5.7. **Lemma.** *Let* $0 \neq D \in \text{Im } F^{s \times s}$ *and* $\rho(D) = \frac{1}{2}(\text{rank } D)(\dim_R F)$. *Recall the Heisenberg group* $H_f = H_{f_D}$ *of* (4.6)-(4.9). *There are integers* $a, b \geqslant 0$ *with* $a+b = \rho(D)(t+u)$, *and an isomorphism* $H_f \to N_{1;a,b}(C)$ *that carries (the action of)* $U(t,u;F)$ *to (the action of) a subgroup of* $U(a,b)$.

Proof. First suppose $F \neq R$. Let $\{1,i\}$ or $\{1,i,j,k\}$ be the standard basis of F over R. There is an element $\gamma \in U(s;F) \subset GL(s;F)$ such that $\gamma D \gamma^{-1}$ has form $\begin{bmatrix} iD' & 0 \\ 0 & 0 \end{bmatrix}$ where D' is a nonsingular real diagonal matrix of size $w = \text{rank } D$. Conjugating by γ, we may assume $D = \begin{bmatrix} iD' & 0 \\ 0 & 0 \end{bmatrix}$, and then $D \cdot F^{s \times (t,u)}$ consists of all $s \times (t+u)$ matrices over F whose last $s-w$ rows are zero. Thus

$$\tau(Z_1,Z_2) = \langle \mathcal{H}(Z_1,Z_2), -iD \rangle + i\langle \mathcal{H}(Z_1,Z_2), D \rangle$$

is a nondegenerate C-hermitian-bilinear form on the underlying complex vector space of $D \cdot F^{s \times (t,u)}$. If π and ν are the number of positive and negative entries in the real diagonal matrix D', then τ has signature (plusses,

minuses) $= (a,b) = ((\dim_{C}F)(t\pi+u\nu), \ (\dim_{C}F)(t\nu+u\pi))$. That gives us a
C-linear isometry $\beta_1: (D\cdot F^{s\times(t,u)},\tau) \to C^{a,b}$.

If $g \in U(t,u;F)$ then $\mathcal{H}(Z_1 g, Z_2 g) = \mathcal{H}(Z_1, Z_2)$, so $\tau(Z_1 g, Z_2 g) = \tau(Z_1, Z_2)$.
Thus β_1 carries $U(t,u;F)$ to a subgroup of $U(a,b)$. Now define

$$\beta_2: \text{Im } F^{s\times s} + D\cdot F^{s\times(t,u)} \to \text{Im } C + C^{a,b} = N_{1;a,b}(C)$$

by

$$\beta_2(D_0,Z_0) = (i<D_0,D>, \ \beta_1(Z_0)).$$

We claim that β_2 is a Lie group homomorphism. In effect, the domain of
β_2 is a closed subgroup of $N_{s;t,u}(F)$, β_2 maps onto $N_{1;a,b}(C)$, and

$$\beta_2(D_1,Z_1)\cdot\beta_2(D_2,Z_2) = (i<D_1 + D_2,D> + \text{Im } \tau(Z_1,Z_2),\beta_1(Z_1 + Z_2))$$

$$= (i<D_1 + D_2 + \mathcal{H}(Z_1,Z_2),D>,\beta_1(Z_1 + Z_2))$$

$$= (i<D_1 + D_2 + \text{Im } \mathcal{H}(Z_1,Z_2),D>,\beta_1(Z_1 + Z_2))$$

$$= \beta_2\{(D_1,Z_1)\cdot(D_2,Z_2)\}.$$

Finally, a glance back at (4.6) shows that β_2 induces the desired isomorphism $\beta: H_{f_D} \to N_{1;a,b}(C)$.

Next suppose $F = R$. There is an element $\gamma \in O(s) \subset GL(s;R)$ such that
$\gamma D\gamma^{-1}$ has form $\begin{bmatrix} D'' & 0 \\ 0 & 0 \end{bmatrix}$, $D'' = \begin{bmatrix} \delta_1 & & \\ & \ddots & \\ & & \delta_\rho \end{bmatrix}$ where $\rho = \rho(D)$ and $\delta_\ell = \begin{bmatrix} 0 & -a_\ell \\ a_\ell & 0 \end{bmatrix}$
with $a_\ell > 0$. Now we conjugate by γ to assume $D = \begin{bmatrix} D'' & 0 \\ 0 & 0 \end{bmatrix}$ as above, and
then $D\cdot R^{s\times(t,u)}$ consists of the real $s \times (t+u)$ matrices whose last $s - 2\rho$
rows are zero. Denote matrix rows by superscripts and define

$$\alpha_1: D \cdot R^{s \times (t,u)} \to C^{\rho \times (t,u)} \quad \text{by } \alpha_1(Z)^{(\ell)} = Z^{(2\ell-1)} + iZ^{(2\ell)}.$$

This is an R-linear isomorphism, it sends the action of $O(t,u)$ to the action of a subgroup of $U(t,u)$, and it carries the action $Z \to DZ$ of D to

$$\alpha_1(Z) \to \alpha_1(DZ) = \tilde{D} \cdot \alpha_1(Z) \quad \text{where } \tilde{D} = \begin{bmatrix} ia_1 & & \\ & \ddots & \\ & & ia_\rho \end{bmatrix}.$$

Now as in the cases $F \neq R$ we define a hermitian form τ on $C^{\rho \times (t,u)}$ by

$$\tau(\alpha_1 Z_1, \alpha_1 Z_2) = \langle \mathcal{H}(\alpha_1 Z_1, \alpha_1 Z_2), -i\tilde{D} \rangle + i\langle \mathcal{H}(\alpha_1 Z_1, \alpha_1 Z_2), \tilde{D} \rangle$$

and we consider the corresponding linear isometry $\beta_1: (C^{\rho \times (t,u)}, \tau) \to C^{\rho t, \rho u}$, thus obtaining a surjective Lie group homomorphism

$$\beta_2: \text{Im } C^{\rho \times \rho} + C^{\rho \times (t,u)} \to N_{1;\rho t, \rho u}(C) \quad \text{by } \beta_2(\tilde{D}_0, \alpha_1 Z_0) = (i\langle \tilde{D}_0, \tilde{D} \rangle, \beta_1 \alpha_1 Z_0).$$

Now we have a surjective Lie group homomorphism

$$\alpha_2: \text{Im } R^{s \times s} + D \cdot R^{s \times (t,u)} \to N_{1;\rho t, \rho u}(C) \quad \text{by } \alpha_2(D_0, Z_0) = (\tfrac{i}{2}\langle D_0, D \rangle, \beta_1 \alpha_1 Z_0)$$

and a glance back at (4.6) shows that it induces the desired isomorpism $\alpha: H_{f_D} \to N_{1;\rho t, \rho u}(C)$.

q.e.d.

Using the isomorphism of Lemma 5.7 and the method of [30, §4] we now prove

___5.8. Proposition.___ *Let* $D \in \text{Im } F^{s \times s}$, $Z \in F^{s \times (t,u)}$ *with* $DZ = 0$, *and* $[\pi_{D,Z}] = [\pi_{f_{D,Z}}] \in N_{s;t,u}(F)^\wedge$ *as in Theorem* 4.14. *Let* $L_{D,Z}$ *be the* $U(t,u;F)$ - *stabilizer of* $[\pi_{D,Z}]$, *described in Lemma* 5.4 *and Proposition* 5.6.

Then $\pi_{D,Z}$ _extends to a unitary representation_ $\tilde{\pi}_{D,Z}$ _of_ $N_{s;t,u}(F) \cdot L_{D,Z}$ _on the representation space of_ $\pi_{D,Z}$.

Proof. Denote $N = N_{s;t,u}(F)$ and $f = f_{D,Z}$ as before. If $D = 0$ then, in (4.6), $\mathfrak{d}_f = \operatorname{Im} F^{s \times s}$ so $\mathfrak{h}_f = 0$ and $\mathfrak{m}_f = \alpha_f = \mathfrak{n}/\operatorname{Im} F^{s \times s}$, commutative. Thus $\pi_{D,Z}$ is the unitary character of N,

$$\pi_{D,Z}(D',Z') = e^{i \operatorname{trace} \operatorname{Re} \mathcal{H}(Z',Z)} ,$$

and it extends to a function $\tilde{\pi}_{D,Z}: N \cdot L_{D,Z} \to \mathbb{C}$ by the formula

$$\tilde{\pi}_{D,Z}((D',Z'),g) = \pi_{D,Z}(D',Z').$$

Using $\mathcal{H}(Z_1 + Z_2 g_1^*, Z) = \mathcal{H}(Z_1,Z) + \mathcal{H}(Z_2,Z)$, from (5.3b), one calculates

$$\tilde{\pi}_{D,Z}((D_1,Z_1),g_1) \cdot \tilde{\pi}_{D,Z}((D_2,Z_2),g_2) = \tilde{\pi}_{D,Z}\{((D_1,Z_1)g_1)((D_2,Z_2),g_2)\},$$

so $\tilde{\pi}_{D,Z}$ is a unitary character.

If $D \neq 0$ but $t + u = 0$, then N is the additive group of $\operatorname{Im} F^{s \times s}$ and $\pi_{D,Z}$ is its unitary character $D' \to e^{i \operatorname{trace} \operatorname{Re} D'D^*}$. Here $L_{D,Z} = \{1\}$ so there is nothing to prove.

Now let $D \neq 0$ with $t + u > 0$, so H_f is a genuine Heisenberg group. Let $\beta: H_f \to N_{1;a,b}(\mathbb{C}) = \operatorname{Im} \mathbb{C} + \mathbb{C}^{a,b}$ be the isomorphism of Lemma 5.7. The central character

$$e^{if_D}: \exp(\mathfrak{d}_f + D') \to e^{if_D(D')} = e^{i \operatorname{trace} \operatorname{Re} D'D^*}$$

of H_f, carries over to a central character

$$\chi_\ell: \operatorname{Im} \mathbb{C} \to \mathbb{C} \quad \text{by} \quad \chi_\ell(w) = e^{\ell w}, \quad 0 \neq \ell = \ell(D) \in \mathbb{R},$$

of $N_{1;a,b}(C)$. The main point of $[30, \S 4]$ was our construction of an
irreducible unitary representation, which I will call ζ_ℓ here, of
$N_{1;a,b}(C)$, with central character χ_ℓ, in such a way that ζ_ℓ extends to a
unitary representation $\tilde{\zeta}_\ell$ of $N_{1,a,b}(C) \cdot U(a,b)$ on the same Hilbert space.
The point of Lemma 5.7 is that the isomorphism β extends to a homomorphism
$\tilde{\beta}: H_f \cdot L_{D,Z} \to N_{1;a,b}(C) \cdot U(a,b)$. Now $\eta_f = \zeta_\ell \cdot \beta$ is an irreducible unitary
representation of H_f with nontrivial character $e^{if_D} = \chi_\ell \cdot \beta$, and it
extends to a unitary representation $\tilde{\eta}_f = \tilde{\zeta}_\ell \cdot \tilde{\beta}$ of $H_f \cdot L_{D,Z}$ on the same
Hilbert space.

Recall the other constituent of $[\pi_{D,Z}] \in \hat{N}$, the unitary character
$\alpha_f = e^{ig_Z}: \exp(\delta_f + Z') \to e^{i \text{ trace Re } \mathcal{H}(Z', Z)}$ on A_f. $L_{D,Z}$ fixes α_f
by (5.3b) so α_f extends to a unitary character on $A_f \cdot L_{D,Z}$ by
$\tilde{\alpha}_f(\exp(\delta_f + Z'),g) = \alpha_f(\exp(\delta_f + Z'))$. Recall the projection
$p: N \to M_f = H_f \times A_f$ and let $\tilde{p}: N \cdot L_{D,Z} \to (H_f \times A_f) \cdot L_{D,Z}$ denote the
obvious extension. Now $\pi_{D,Z} = (\eta_f \otimes \alpha_f) \cdot p$ belongs to the unitary representation class $[\pi_{D,Z}] \in \hat{N}$ for the coadjoint orbit $\text{Ad}^*(N) \cdot f_{D,Z}$, and it
extends to a unitary representation $\tilde{\pi}_{D,Z} = (\tilde{\eta}_f \otimes \tilde{\alpha}_f) \cdot \tilde{p}$ of $N \cdot L_{D,Z}$ on the
same Hilbert space. Were $\pi'_{D,Z}$ another representative of $[\pi_{D,Z}]$, its
equivalence with $\pi_{D,Z}$ would pull $\tilde{\pi}_{D,Z}$ back to an extension of $\pi'_{D,Z}$
from N to $N \cdot L_{D,Z}$.

$$q.e.d.$$

We now have all the information needed to go from $N_{s;t,u}(F)^{\hat{}}$ to
$G_{s;t,u}(F)^{\hat{}}$ by the Mackey little-group method. It fits together as follows.
Let $D \in \text{Im } F^{s \times s}$ and $Z \in F^{s \times (t,u)}$ with $DZ = 0$. The corresponding unitary
representation class $[\pi_{D,Z}] \in N_{s;t,u}(F)^{\hat{}}$ has $U(t,u;F)$-stabilizer
$L_{D,Z} = \{g \in U(t,u;F): Zg = Z\}$, and $[\pi_{D,Z}]$ extends to a unitary representation class $[\tilde{\pi}_{D,Z}] \in \{N_{s;t,u}(F) \cdot L_{D,Z}\}^{\hat{}}$. The associated unitary representation classes for $G_{s;t,u}(F) = N_{s;t,u}(F) \cdot U(t,u;F)$ are the induced classes

(5.9a) $[\pi_{D,Z,\gamma}] = [\text{Ind}_{N_{s;t,u}(F) \cdot L_{D,Z} \uparrow G_{s;t,u}(F)} (\tilde{\pi}_{D,Z} \otimes \gamma)]$

where

(5.9b) $[\gamma] \in \hat{L}_{D,Z}$ extended to $N_{s;t,u}(F) \cdot L_{D,Z}$ by $\gamma((D',Z'),g) = \gamma(g)$.

In view of Theorem 4.14, every irreducible unitary representation of $G_{s;t,u}(F)$ is equivalent to one of these $\pi_{D,Z,\gamma}$.

Let us denote the family of representations arising from $[\pi_{D,Z}]$ by

(5.10a) $G_{s;t,u}(F)\hat{}_{D,Z} = \{[\pi_{D,Z,\gamma}]: [\gamma] \in \hat{L}_{D,Z}\}.$

According to the equivalence criterion (A.10) of the little group method,

(5.10b) $[\pi_{D,Z,\gamma}] = [\pi_{D',Z',\gamma'}] \Longleftrightarrow \begin{Bmatrix} D' = D, \ Z' \in Z \cdot U(t,u;F), \ \text{and} \\ \gamma' \ \text{is suitably related to} \ \gamma \end{Bmatrix}.$

In particular, families

(5.10c) $G_{s;t,u}(F)\hat{}_{D,Z}$ and $G_{s;t,u}(F)\hat{}_{D',Z'}$ are

                    equal if $D' = D$, $Z' \in Z \cdot U(t,u;F)$,
                    disjoint otherwise   .

To be explicit about the partitioning (5.10) of $G_{s;t,u}(F)\hat{}$ we glance at the linear subspace S_Z of $F^{t,u}$ spanned by the columns of Z^*. First, Witt's Theorem tells us

(5.11a) $Z' \in Z \cdot U(t,u;F) \Longleftrightarrow \text{rank } Z' = \text{rank } Z$ and $\mathcal{H}(Z',Z') = \mathcal{H}(Z,Z)$.

Second, the integers a, b, c associated to S_Z as in Proposition 5.6 by

$$c = \dim(S_Z \cap S_Z^\perp) \quad \text{and} \quad S_Z \cong (S_Z \cap S_Z^\perp) \oplus F^{a,b}$$

are specified by the rank Z together with the "signature"

$$(\text{\# positive eigenvalues, \# negative eigenvalues, nullity})$$

of the hermitian matrix $\mathcal{H}(Z,Z)$. Here recall that every $s \times s$ hermitian matrix H over F can be diagonalized, $UHU^{-1} = \text{diag }\{a_1,\ldots,a_s\}$ with $U \in U(s;F)$ and a_ℓ real, so this "signature" has a meaning. Now

(5.11b) $\mathcal{H}(Z,Z)$ has "signature" $(a,b,c + \{s - \text{rank } Z\})$.

Using Proposition 5.6, we formulate the preceding results and discussion as follows.

 5.12. Theorem. $G_{s;t,u}(F)^\wedge$ *is the disjoint union of non-empty subsets* $S(a,b,c;D)$, $D \in \text{Im } F^{s\times s}$ *and* a,b,c *non-negative integers such that*

(*) $a + b + c + \text{rank } D \leqslant s, \quad a + c \leqslant t, \quad b + c \leqslant u,$

given as follows. Let $\mathcal{H}_{a,b;D}$ *denote the space of all* $s \times s$ *hermitian matrices* H *over* F *such that* $DH = 0$ *and* H *has "signature"* $(a,b,s-a-b)$. *Then* $S(a,b,c;D)$ *is parameterized by*

$$\mathcal{H}_{a,b;D} \times G_{c;t-c-a,u-c-b}(F)^\wedge$$

under

$$(H,[\gamma]) \leftrightarrow [\pi_{D,Z,\gamma}]$$

where

$$Z \in F^{s \times (t,u)} \quad \underline{with} \quad DZ = 0, \quad \mathcal{H}(Z,Z) = H \quad \underline{and} \quad \text{rank } Z = a+b+c$$

and

$$L_{D,Z} \quad \underline{is\ identified\ with\ its\ isomorph} \quad G_{c;t-c-a,u-c-b}(F).$$

In other words, $S(a,b,c;D)$ *is the union of the* $G_{s;t,u}(F)^{\wedge}_{D,Z}$ *such that* $Z \in F^{s \times (t,u)}$ *with* $DZ = 0$, rank $Z = a+b+c$ *and* $\mathcal{H}(Z,Z)$ *of "signature"* $(a,b,s-a-b)$.

Proof. Let $S(a,b,c;D)$ denote the union of the $G_{s;t,u}(F)^{\wedge}_{D,Z}$ such that $Z \in F^{s \times (t,u)}$ with $DZ = 0$, rank $Z = a+b+c$ and $\mathcal{H}(Z,Z)$ of "signature" $(a,b,s-a-b)$. It only remains to show that $S(a,b,c;D)$ is non-empty exactly when

$$(*) \qquad a + b + c + \text{rank } D \leqslant s, \quad a + c \leqslant t \quad \text{and} \quad b + c \leqslant u,$$

to show that $G_{s;t,u}(F)^{\wedge}$ is the disjoint union of the nonempty $S(a,b,c;D)$, and to verify the parameterization.

Let $[\pi] \in G_{s;t,u}(F)^{\wedge}$. Then $[\pi] = [\pi_{D,Z,\gamma}]$ where $D \in \text{Im } F^{s \times s}$, $Z \in F^{s \times (t,u)}$ with $DZ = 0$ and $\gamma \in \hat{L}_{D,Z}$. There is no freedom of choice for D, but the choice of Z is free so long as rank Z and $\mathcal{H}(Z,Z)$ are not altered. Thus $[\pi] = [\pi_{D,Z,\gamma}]$ determines integers a, b, $c \geqslant 0$ by: $\mathcal{H}(Z,Z)$ has "signature" $(a,b,s-a-b)$ and Z has rank $a+b+c$. In other words, we have checked that $[\pi_{D,Z,\gamma}] \rightarrow (\mathcal{H}(Z,Z), [\gamma])$ is a one to one map from $S(a,b,c;D)$ into $\mathcal{H}_{a,b;D} \times G_{c;t-c-a,u-c-b}(F)^{\wedge}$, and that $G_{s;t,u}(F)^{\wedge}$ is the disjoint union of the non-empty $S(a,b,c;D)$.

Given $[\pi] = [\pi_{D,Z,\gamma}]$ as above, the integers a,b,c are characterized in terms of the subspace S_Z of $F^{t,u}$ spanned by the columns of Z^*:

$$c = \dim(S_Z \cap S_Z^{\perp}) \quad \text{and} \quad S_Z \cong (S_Z \cap S_Z^{\perp}) \oplus F^{a,b}.$$

It follows immediately that $a + c \leqslant t$ and $b + c \leqslant u$. $DZ = 0$ implies rank $Z \leqslant$ nullity $D = s -$ rank D, so also $a + b + c +$ rank $D =$ rank $Z +$ rank $D \leqslant s$. Thus (*) is a necessary condition for $S(a,b,c;D)$ to be non-empty.

Conversely, fix $D \in \text{Im } F^{s \times s}$ and let a, b and c be non-negative integers that satisfy (*). Choose $\gamma \in U(s;F)$ with $\gamma D \gamma^{-1} = \begin{pmatrix} D' & 0 \\ 0 & 0 \end{pmatrix}$ where D' is a nonsingular skew hermitian $r \times r$ matrix over F, $r =$ rank D. If $H \in \mathcal{H}_{a,b,D}$ now $DH = 0$ says $\gamma D \gamma^{-1} \cdot \gamma H \gamma^{-1} = 0$ so $\gamma H \gamma^{-1} = \begin{pmatrix} 0 & 0 \\ 0 & H' \end{pmatrix}$ where H' is an $(s-r) \times (s-r)$ hermitian matrix over F with "signature" $(a,b,s-r-a-b)$. Because of (*), $F^{t,u}$ has a subspace S such that $c = \dim(S \cap S^{\perp})$ and $S \cong (S \cap S^{\perp}) \oplus F^{a,b}$. S is (over)-spanned by $s-r$ vectors $\{z_1, \ldots, z_{s-r}\}$ such that $H' = (h(z_{\ell}, z_m))$. Let Z' be the $(s-r) \times (t+u)$ matrix where ℓ-th row is the transpose of z_{ℓ} and define $Z = \gamma^{-1} \begin{pmatrix} 0 \\ Z' \end{pmatrix} \in F^{s \times (t,u)}$. Then

$$DZ = \gamma^{-1} \begin{pmatrix} D' & 0 \\ 0 & 0 \end{pmatrix} \gamma \cdot \gamma^{-1} \begin{pmatrix} 0 \\ Z' \end{pmatrix} = 0$$

and

$$\mathcal{H}(Z,Z) = \mathcal{H}\left(\gamma^{-1} \begin{pmatrix} 0 \\ Z' \end{pmatrix}, \gamma^{-1} \begin{pmatrix} 0 \\ Z' \end{pmatrix}\right) = \gamma^{-1} \begin{pmatrix} 0 & 0 \\ 0 & H' \end{pmatrix} \gamma = H.$$

As rank $Z = \dim S = a + b + c$, now $[\pi_{D,Z,\gamma}]$ exists, belongs to $S(a,b,c;D)$, and corresponds to the parameter value $(H, [\gamma])$ there. Thus (*) is a sufficient condition for $S(a,b,c;D)$ to be non-empty, and $[\pi_{D,Z,\gamma}] \to (\mathcal{H}(Z,Z), [\gamma])$ maps $S(a,b,c;D)$ onto $\mathcal{H}_{a,b;D} \times G_{c;t-c-a,u-c-b}(F)^{\wedge}$ when (*) holds.

We have shown that (*) is a necessary and sufficient condition that $S(a,b,c;D)$ be non-empty, and that, given (*), the set $S(a,b,c;D)$ is bijectively parameterized by $\mathcal{H}_{a,b;D} \times G_{c;t-c-a,u-c-b}(F)^{\wedge}$. That completes the proof of Theorem 5.12.

                                                            q.e.d.

5.13. Remark. Theorem 5.12 gives a complete description of $G_{s;t,u}(F)^{\wedge}$ soon as one knows the $U(\ell,m;F)^{\wedge}$ for $0 \leqslant \ell \leqslant t$ and $0 \leqslant m \leqslant u$. In effect, the case $a = b = c = 0$ gives $\mathcal{H}_{0,0;D} = \{0\}$ and $G_{0;t,u}(F) = N_{0;t,u}(F) \cdot U(t,u;F) = U(t,u;F)$, so $S(0,0,0;D)$ consists of the classes

$$[\pi_{D,0,\gamma}] = [\tilde{\pi}_{D,0} \otimes \gamma] \quad \text{with} \quad [\gamma] \in U(t,u;F)^{\wedge} .$$

If a, b and c are not all zero, then either $t-c-a < t$ or $u-c-b < u$, by induction on $t + u$ we know $G_{c;t-c-a,u-c-b}(F)^{\wedge}$ as soon as we know the $U(\ell,m;F)^{\wedge}$ for $0 \leqslant \ell \leqslant t-c-a$ and $0 \leqslant m \leqslant u-c-b$, and by Theorem 5.12 we know $S(a,b,c;D)$ as soon as we know $G_{c;t-c-a,u-c-b}(F)^{\wedge}$.

The $U(\ell,m;F)$, $0 \leqslant \ell \leqslant t$ and $0 \leqslant m \leqslant u$, are well-known classical reductive Lie groups. One has explicit knowledge of the various series in $U(\ell,m;F)^{\wedge}$ that contribute to Plancherel measure, for example through the work of Harish-Chandra, and this will suffice when we write down the Plancherel formula for $G_{s;t,u}(F)$. The work of Langlands [3] is close to a complete description of the $U(\ell,m;F)^{\wedge}$, but he works with Banach representations and the unitarization problem remains. Finally, the $U(\ell,m;F)^{\wedge}$ are known for some small values of ℓ and m through the work of the Moscow School.

§6. Representations of the Maximal Parabolic Subgroups

We now combine the results of §§1 and 5 with the Mackey little-group method and find the irreducible unitary representations of the maximal parabolic subgroups of the unitary groups $U(p,q;F)$. Those parabolic subgroups are the

$$P_E = \{g \in U(p,q;F): gE = E\}$$

where E is a totally isotropic subspace of dimension $s > 0$ in $F^{p,q}$. According to Theorem 2.10, P_E is isomorphic to

$$P_{s;p-s,q-s}(F) = G_{s;p-s,q-s}(F) \cdot GL(s;F), \quad 1 \leq s \leq \min(p,q).$$

The procedure for determining the unitary dual of $P_{s;p-s,q-s}(F)$ is valid for its maximal unimodular subgroup $P'_{s;p-s,q-s}(F) = G_{s;p-s,q-s}(F) \cdot GL'(s;F)$ and also for various classes of subgroups in which $GL(s;F)$ is cut down in a reasonable manner.

For convenience in calculation and in reference to §5, we set $t = p-s$ and $u = q-s$.

   6.1. Lemma. *The* $GL(s;F)$-*stabilizer of the class* $[\pi_{D,Z,\gamma}] \in G_{s;t,u}(F)^\wedge$ *defined in* (5.9) *is*

$$M_{D,Z} = \{\beta \in GL(s;F): \beta^* D\beta = D \;\; \underline{and} \;\; \beta^* Z - Zg \in D \cdot F^{s \times (t,u)} \;\; \underline{where} \quad g \in U(t,u;F)\}.$$

Let $d = \operatorname{rank} D$. *If* $F = C$ *let* $(d',d'',s-d)$ *denote the "signature" of* $i^{-1}D$. *Recall* $\mathcal{H}$ *from* (5.11) *and let* $(a,b,s-a-b)$ *denote the "signature" of* $\mathcal{H}(Z,Z)$. *Then*

(6.2a) $\qquad M_{D,Z} \cong F^{d \times (s-d)} \cdot \{H \times K\}$ _where_

(6.2b) $\qquad K = \{(\chi; \sigma, \tau) \in L_{s-d-a-b, a+b}(F) : \tau \in U(a,b;F)\}$,

(6.2c) $\qquad$ _if_ $F = R$ _then_ $H = Sp(d/2;R)$, _defined in_ (6.4) _below,_

(6.2d) $\qquad$ _if_ $F = C$ _then_ $H = U(d', d'')$, _and_

(6.2e) $\qquad$ _if_ $F = Q$ _then_ $H = SO^*(2d)$, _defined in_ (6.5) _below._

{_In particular,_ $K \cong P_{s-d-a-b; a,b}(F)/\mathrm{Im}\, F^{(s-d-a-b) \times (s-d-a-b)}$}.

Proof. Since $GL(s;F)$ centralizes $U(t,u;F)$ in $P_{s;t,u}(F)$, we combine Lemma 5.2 with Proposition 4.3 and the construction (5.9) to see

$$M_{D,Z} = \{\beta \in GL(s;F): \beta^* D \beta = D \text{ and } \beta^* Z - Zg \in D \cdot F^{s \times (t,u)} \quad \text{for some}$$

$g \in U(t,u;F)\}$, as asserted.

We replace $[\pi_{D,Z,\gamma}]$ by a $GL(s;F)$-conjugate $[\pi_{D,Z,\gamma} \cdot \beta^{-1}]$, thus replacing D by $(\beta^*)^{-1} D \beta^{-1}$. Choosing β appropriately, this means that we assume $D = \begin{pmatrix} D' & 0 \\ 0 & 0 \end{pmatrix}$ where $D' \in \mathrm{Im}\, F^{d \times d}$ is the nonsingular matrix given by

(6.3) $D' = iI$ if $F = Q$, $D' = i\begin{pmatrix} I_{d'} & 0 \\ 0 & -I_{d''} \end{pmatrix}$ if $F = C$, $D' = \begin{pmatrix} 0 & I \\ -I & 0 \end{pmatrix}$ if $F = R$.

Now express an arbitrary element $\beta \in GL(s;F)$ in block form $\begin{pmatrix} a & b \\ c & d \end{pmatrix}$ and calculate $\beta^* D \beta$ to see that

$$\beta^* D \beta = D \Leftrightarrow a^* D' a = D' \quad \text{and} \quad b = 0.$$

$DZ = 0$ is equivalent to $Z = \begin{pmatrix} 0 \\ Z' \end{pmatrix}$ with $Z' \in F^{(s-d) \times (t,u)}$ because $D = \begin{pmatrix} D' & 0 \\ 0 & 0 \end{pmatrix}$ with D' nonsingular. For the same reason, $D \cdot F^{s \times (t,u)}$ consists of all $\begin{pmatrix} Z'' \\ 0 \end{pmatrix}$ with $Z'' \in F^{d \times (t,u)}$. Assuming $\beta^* D \beta = D$ we calculate

$$\beta^* Z - Zg = \begin{pmatrix} c^*Z' \\ d^*Z' - Z'g \end{pmatrix}; \quad \text{so} \quad \beta^* Z - Zg \in D \cdot F^{s \times (t,u)} \Leftrightarrow d^*Z' = Z'g.$$

Now we have

$$M_{D,Z} \cong \left\{ \begin{pmatrix} a & 0 \\ c & d \end{pmatrix} : \ a^*D'a = D' \quad \text{and} \quad d^*Z' \in Z' \cdot U(t,u;F) \right\}.$$

The automorphism $\beta \to (\beta^*)^{-1}$ of $GL(s;F)$ carries this to

$$M_{D,Z} \cong \left\{ \begin{pmatrix} a & b \\ 0 & d \end{pmatrix} : \ aD'a^* = D' \quad \text{and} \quad dZ' \in Z' \cdot U(t,u;F) \right\}.$$

Now the isomorphism of Lemma 1.4 gives $M_{D,Z} \cong F^{d \times (s-d)} \cdot \{H \times K\}$ where

$$H = \{a \in GL(d;F): aD'a^* = D'\} \quad \text{and} \quad K = \{d \in GL(s-d;F): dZ' \in Z' \cdot U(t,u;F)\}.$$

From (5.11a) $K = \{d \in GL(s-d;F): d \cdot \mathcal{H}(Z',Z') \cdot d^* = \mathcal{H}(Z',Z')\}$. If the hermitian matrix $\mathcal{H}(Z',Z')$ has "signature" $(a,b,s-d-a-b)$, then we split $F^{s-d} = E' \oplus F^{a,b}$ where E' is its null space. Following the isomorphism of Lemma 1.4, this gives us

$$K \cong \left\{ \begin{pmatrix} \sigma & \chi\tau \\ 0 & \tau \end{pmatrix} : \ \sigma \in GL(s-d-a-b;F) \quad \text{and} \quad \tau \in U(a,b;F) \right\}.$$

In other words, in the notation (1.3), $K \cong \{(\chi,\sigma,\tau) \in L_{s-d-a-b,a+b}(F): \tau \in U(a,b;F)\}$, as required.

If $F = R$ then (6.3) shows d even and $H = Sp(d/2;R)$, real symplectic group, defined as follows. Let α be the nondegenerate antisymmetric bilinear form on R^{2v} given by

$$(6.4a) \qquad\qquad \alpha(x,y) = \sum_{1 \leq \ell \leq v} (x^\ell y^{v+\ell} - x^{v+\ell} y^\ell).$$

Then the real symplectic group is

(6.4b) $Sp(v;R) = \{a \in GL(2v;R): \alpha(ax,ay) = \alpha(x,y), \text{ all } x, y \in R^{2v}\}.$

If $F = C$ then the defining condition $aD'a^* = D'$ of H is, by (6.3), equivalent to the defining condition $a \cdot \begin{bmatrix} I_{d'} & 0 \\ 0 & -I_{d''} \end{bmatrix} \cdot a^* = \begin{bmatrix} I_{d'} & 0 \\ 0 & -I_{d''} \end{bmatrix}$ of the

indefinite complex unitary group $U(d',d'') = U(d',d'';C).$

If $F = Q$ then (6.3) $D' = iI$ where $\{1,i,j,k\}$ is the standard basis of Q over R. We take $C = R + iR \subset Q$. Observe that H has Lie algebra

$$\mathfrak{h} = \{\alpha \in Q^{d \times d}: \alpha i + i\alpha^* = 0\}.$$

If $\alpha \in Q^{d \times d}$ we decompose it as $\alpha = \sigma + j\tau$ with $\sigma, \tau \in C^{d \times d}$. Then

$$\alpha i + i\alpha^* = \sigma i + j\tau i + i\sigma^* - i\tau^* j = (\sigma + \sigma^*)i + j(\tau + {}^t\tau)i$$

so

$$\mathfrak{h} = \mathfrak{k} + \mathfrak{p} \text{ where } \mathfrak{k} = \{\sigma \in C^{d \times d}: \sigma + \sigma^* = 0\} \text{ and } \mathfrak{p} = \{j\tau \in C^{d \times d}: \tau + {}^t\tau = 0\}.$$

This is the eigenspace decomposition of $\mathfrak{h}$ under the Cartan involution $\beta \to -\beta^*$ of the Lie algebra $\mathfrak{gl}(d;Q)$. The 1-eigenspace, $\mathfrak{k}$, is the Lie algebra $\mathfrak{u}(d)$ of the unitary group $U(d) \subset GL(d;C)$. The action of $\mathfrak{k}$ on the (-1)-eigenspace, $\mathfrak{p}$, is the exterior square of the dual of its usual (vector) representation on C^d. Denote

(6.5a) $SO(2d;C):$ complex special orthogonal group

and

(6.5b) $SO^*(2d):$ its real form with maximal compact subgroup $U(d).$

We have shown that H is locally isomorphic to $SO^*(2d)$. Since the Cartan involution $a \to iai^{-1}$ of H has fixed point set $H \cap GL(d;C) = U(d)$, the isomorphism is global.

<u>q.e.d.</u>

    <u>6.6. Lemma</u>. If $[\pi_{D,Z,\gamma}] \in G_{s;t,u}(F)^{\wedge}$, then $\pi_{D,Z,\gamma}$ <u>extends</u> <u>to</u> <u>a</u> <u>unitary</u> <u>representation</u> $\tilde{\pi}_{D,Z,\gamma}$ <u>of</u> $G_{s;t,u}(F) \cdot M_{D,Z}$ <u>on</u> <u>the</u> <u>same</u> <u>Hilbert</u> <u>space</u>.

    <u>Proof</u>. The stabilizer of $f_{D,Z}$ in $U(t,u;F) \times GL(s;F)$ is $Q_{D,Z} = \{(g,\beta): \beta^* D\beta = D \text{ and } \beta^* Zg = Z\}$. Following the proof of Proposition 5.8, one sees that $\pi_{D,Z}$ extends to a unitary representation $\pi'_{D,Z}$ of $N_{s;t,u}(F) \cdot Q_{D,Z}$ on the same Hilbert space.

    Let $R_{D,Z}$ denote the projection of $Q_{D,Z}$ to $U(t,u;F)$. In the notation of Lemma 5.4, $R_{D,Z} = \{g \in U(t,u;F): gS_Z = S_Z\}$, and $L_{D,Z}$ is the subgroup $g|_{S_Z} = $ identity. Let V_γ denote the $R_{D,Z}$ -stabilizer of $[\gamma] \in \hat{L}_{D,Z}$. By induction on s , $[\gamma]$ extends to a class $[\tilde{\gamma}] \in \hat{V}_\gamma$.

    Now $[\tilde{\pi}_{D,Z} \otimes \gamma]$ extends from $N_{s;t,u}(F) \cdot L_{D,Z}$ to an irreducible unitary equivalence class $[\omega]$ on $W = N_{s;t,u}(F) \cdot \{Q_{D,Z} \cap (V_\gamma \times GL(s;F))\}$ by the formula $\omega(D',Z';g,\beta) = \pi'_{D,Z}(D',Z';g,\beta) \otimes \tilde{\gamma}(g)$. Define

$$\tilde{\pi}_{D,Z,\gamma} = \mathrm{Ind}_{W \uparrow G_{s;t,u}(F) \cdot M_{D,Z}} (\omega).$$

As $Q_{D,Z} \cap (U(t,u;F) \times 1) = L_{D,Z}$, we have $W \cap G_{s;t,u}(F) = N_{s;t,u}(F) \cdot L_{D,Z}$. Thus the fibration used in the induction that defines $\tilde{\pi}_{D,Z,\gamma}$, has base space

$$G_{s;t,u}(F) \cdot M_{D,Z}/W \approx G_{s;t,u}(F)/N_{s;t,u}(F) \cdot L_{D,Z}.$$

Thus the restriction of $\tilde{\pi}_{D,Z,\gamma}$ to $G_{s;t,u}(F)$ is

$$\mathrm{Ind}_{N_{s;t,u}(F) \cdot L_{D,Z}} \uparrow G_{s;t,u}(F) \, ({}^{(\omega}|_{N_{s;t,u}(F) \cdot L_{D,Z}}) =$$

$$= \mathrm{Ind}_{N_{s;t,u}(F) \cdot L_{D,Z}} \uparrow G_{s;t,u}(F) \, (\tilde{\pi}_{D,Z} \otimes \gamma),$$

which is $\pi_{D,Z,\gamma}$, as required.

q.e.d.

Mackey's little-group method now says that $P_{s;t,u}(F)\hat{\,}$ consists of the unitarily induced classes

(6.7a) $[\pi_{D,Z,\gamma,\mu}] = [\mathrm{Ind}_{G_{s;t,u}(F) \cdot M_{D,Z}} \uparrow P_{s;t,u}(F) \, (\tilde{\pi}_{D,Z,\gamma} \otimes \mu)]$

where

(6.7b) $D \in \mathrm{Im} \, F^{s \times s}, \; Z \in F^{s \times (t,u)}$ with $DZ = 0, \; [\gamma] \in \hat{L}_{D,Z}$

and

(6.7c) $[\mu] \in \hat{M}_{D,Z}$ is extended by $\mu(D',Z';g,\beta) = \mu(\beta).$

Classes $[\pi_{D,Z,\gamma,\mu}] = [\pi_{D',Z',\gamma',\mu'}]$ if and only if there is an element $(g,\beta) \in U(t,u;F) \times GL(s;F)$ such that

(6.8a) $D' = \beta^* D \beta$ and $Z' - \beta^* Z g \in D \cdot F^{s \times (t,u)},$

(6.8b) $g' \to \gamma(g^{-1} g' g)$ is equivalent to γ', and

(6.8c) $\beta' \to \mu(\beta^{-1} \beta' \beta)$ is equivalent to μ'.

With Lemma 6.1 in mind, we denote

(6.9a) $J_{d;e,a,b}(Q) = Q^{d\times(e+a+b)} \cdot \{SO^*(2d) \times L_{e;a,b}(Q)\},$

(6.9b) $J_{d',d'';e,a,b}(C) = C^{(d',d'')\times(e+a+b)} \cdot \{U(d',d'') \times L_{e;a,b}(C)\},$ and

(6.9c) $J_{d;e,a,b}(R) = R^{d\times(e+a+b)} \cdot \{Sp(d/2;R) \times L_{e;a,b}(R)\}$ for d even.

Here, using (1.3a),

(6.9d) $L_{e;a,b}(F) = \{(\chi;\sigma,\tau) \in L_{e,a+b}(F) : \tau \in U(a,b;F)\}.$

Schematically, $J_{d \text{ or } d',d'';e,a,b}(F)$ consists of the matrices

$$
\left[
\begin{array}{c|c|c}
\multicolumn{1}{c|}{SO^*(2d) \text{ or } U(d',d'') \text{ or } Sp(d/2;R)} & \multicolumn{2}{c}{F^{d\times(e+a+b)}} \\
\hline
\multirow{2}{*}{0} & GL(e;F) & F^{e\times(a+b)} \\
\cline{2-3}
 & 0 & U(a,b;F) \\
\end{array}
\right]
$$

Now we can state and prove the principal result for maximal parabolic sub-groups of unitary groups.

   6.10.   Theorem. $P_{s;t,u}(F)^\wedge$ _is the disjoint union of non-empty subsets_
$S(a,b,c;d)$ _where_

(*) $\begin{cases} F = Q: & a,b,c,d \quad \text{integers} \geq 0, \quad a+b+c+d \leq s, \quad a+c \leq t,\ b+c \leq u \\ F = C: & a,b,c,d',d'' \text{ integers} \geq 0, \quad a+b+c+d'+d'' \leq s,\ a+c \leq t,\ b+c \leq u \\ F = R: & a,b,c,\frac{1}{2}d \quad \text{integers} \geq 0, \quad a+b+c+d \leq s, \quad a+c \leq t,\ b+c \leq u \end{cases}$

_If_ $(a,b,c;d)$ _satisfies_ (*), $d = (d',d'')$ _when_ $F = C$, _define_
$D = \begin{pmatrix} D' & 0 \\ 0 & 0 \end{pmatrix} \in \text{Im } F^{s\times s}$ _by_

$D' = iI$ _if_ $F = Q$, $D' = \begin{pmatrix} I_{d'} & 0 \\ 0 & -I_{d''} \end{pmatrix}$ _if_ $F = C$, $D' = \begin{pmatrix} 0 & I \\ -I & 0 \end{pmatrix}$ _if_ $F = R$.

Choose $Z \in F^{s \times (t,u)}$ _such that_

$DZ = 0$, $\mathcal{H}(Z,Z)$ _has "signature"_ $(a,b,s-a-b)$, _and_ rank $Z = a+b+c$.

Then $S(a,b,c;d)$ _is parameterized by_

$$G_{c;t-c-a,u-c-b}(F)^{\wedge} \times J_{d;s-d-a-b,a,b}(F)^{\wedge}, \quad d = (d',d'') \quad \underline{when} \quad F = C,$$

under

$$([\gamma], [\mu]) \leftrightarrow [\pi_{D,Z,\gamma,\mu}]$$

where

$$L_{D,Z} \quad \underline{is\ identified\ with\ its\ isomorph} \quad G_{c;t-c-a,u-c-b}(F)$$

and

$$M_{D,Z} \quad \underline{is\ identified\ with\ its\ isomorph} \quad J_{d;s-d-a-b;a,b}(F).$$

Note. $G_{c;t-c-a,u-c-b}(F)^{\wedge}$ is described in Theorem 5.12. For continuity of exposition we defer the description of $J_{d;s-d-a-b,a,b}(F)^{\wedge}$.

Proof. Let $[\pi_{D,Z,\gamma,\mu}] \in P_{s;t,u}(F)^{\wedge}$. Conjugating by $\beta \in GL(s;F)$, we send D to $(\beta^*)^{-1}D\beta^{-1}$, and thus may assume $D = \begin{pmatrix} D' & 0 \\ 0 & 0 \end{pmatrix}$ as in (6.3). Now $Z = \begin{pmatrix} 0 \\ Z' \end{pmatrix}$ where, also writing d for $d' + d''$ in the complex case, $Z' \in F^{(s-d) \times (t,u)}$ has rank $a+b+c \leq \min(s-d,t+u)$, and $\mathcal{H}(Z,Z)$ has "signature" $(a,b,s-a-b)$ with $a+c \leq t$ and $b+c \leq u$. These inequalities are restated by (*).

As in the proof of Lemma 6.1, fixing D the matrix Z is still free up to a transformation $Z' \to \beta'Z'g$ with $\beta' \in GL(s-d;F)$ and $g \in U(t,u;F)$. As $\mathcal{H}(\beta'Z'g,\beta'Z'g) = \beta' \cdot \mathcal{H}(Z',Z') \cdot \beta'^*$, this sends $\mathcal{H}(Z,Z) = \begin{pmatrix} 0 & 0 \\ 0 & \mathcal{H}(Z',Z') \end{pmatrix}$ to $\begin{pmatrix} 0 & 0 \\ 0 & \beta' \cdot \mathcal{H}(Z',Z') \cdot \beta'^* \end{pmatrix}$. The only invariants of such $Z' \to \beta'Z'g$ are a,b and c.

$\underline{q.e.d.}$

§7. Representations of the Little-Groups $J_{d;e,a,b}(F)$.

We complete the determination of the unitary duals of the maximal parabolic subgroups $P_{s;p-s,q-s}(F) \subset U(p,q;F)$ given by Theorems 2.10 and 6.10.
This is a matter of describing the unitary duals of the little-groups
$J_{d;e,a,b}(F)$ that occur in Theorem 6.10. The nature of the problem requires
that we look ahead of §§8-10 for the case $F = R$, to §§13-15 for the case
$F = Q$. Thus the reader may wish to defer reading this section until he
glances through §§8, 9, 10, 13, 14 and 15.

Given non-negative integers d, d_i and c we denote

(7.1a) $$I_{d;c}(Q) = Q^{d \times c} \cdot SO^*(2d),$$

(7.1b) $$I_{d;c}(C) = C^{d \times c} \cdot U(d) \quad \text{where} \quad d = (d',d''), \quad \text{and}$$

(7.1c) $$I_{d;c}(R) = R^{d \times c} \cdot Sp(d/2;R) \quad \text{where} \quad d \quad \text{is even.}$$

Glancing back to (6.9), we see that $I_{d;e+a+b}(F)$ is a closed normal subgroup
of $J_{d;e,a,b}(F)$, and that the latter is the semidirect product group

(7.2a) $$J_{d;e,a,b}(F) = I_{d;e+a+b}(F) \cdot L_{e;a,b}(F)$$

where

(7.2b) $$L_{e;a,b}(F) = F^{e \times (a,b)} \cdot \{U(a,b;F) \times GL(e;F) \cong P_{e;a,b}(F)/\text{Im } F^{e \times e}.$$

§7A. The Case $F = R$.

Looking ahead to (8.5) and Theorem 8.6, we see that

(7.3a) $$I_{d;c}(R) \cong G_{c;d}(R)/Sym\ R^{c \times c}$$

where R^{2c+d} is equipped with the nondegenerate antisymmetric bilinear form
(8.1a), E is a fixed totally isotropic c-dimensional subspace, and

(7.3b) $$G_{c;d}(R) \cong \{g \in Sp(c + \tfrac{1}{2}d;R): g|_E = identity\}.$$

Further, Sym $R^{c \times c}$ is central in the unipotent radical $N_{c;d}(R)$ of
$G_{c;d}(R)$ and $N_{c;d}(R)/Sym\ R^{c \times c}$ is the (commutative, additive) group $R^{c \times d}$.
In the notation (9.29) and (9.3), that tells us

(7.4) $$I_{d;c}(R)^{\wedge} = \{[\pi_{D,Z,\gamma}] \in G_{c;d}(R)^{\wedge}:\ D = 0\}\ .$$

In view of (9.20b), Corollary 9.34 specializes to

   7.5. Proposition. $I_{d;c}(R)^{\wedge}$ *is the disjoint union of non-empty subsets*
$S_I(\ell,m)$ *where* ℓ *and* m *are non-negative integers such that*

(*) $$\ell + 2m \leqslant c \quad \text{and} \quad 2(\ell + m) \leqslant d,$$

given as follows. Let $\mathcal{A}_m$ *denote the space of all* $c \times c$ *antisymmetric real*
matrices of rank 2m. *Then* $S_I(\ell,m)$ *is parameterized by*
$\mathcal{A}_m \times G_{\ell;d+2c-2\ell-2m}(R)^{\wedge}$ *under*

          $$(A,[\gamma]) \leftrightarrow [\pi_{0,Z,\gamma}]\ \text{in the notation} \ (7.4)\ \text{and}\ (9.29)$$

where $Z \in R^{c \times d}$ *with* $\mathcal{B}(Z,Z) = A$ *and* rank $Z = \ell + 2m$.

   The $L_{e;a,b}(R)$-stabilizer of $[\pi_{0,Z,\gamma}] \in I_{d;c}(R)^{\wedge}$ is obtained by
specializing Lemma 10.1 to the case $D = 0$. If $Z \in R^{c \times d}$ with rank
$Z = \ell + 2m$ and rank $\mathcal{B}(Z,Z) = 2m$, as in Proposition 7.5, then

$c = e + a + b$ and that stabilizer is

(7.6a) $M_Z = \{\beta \in L_{e;a,b}(R) \subset GL(c;R): {}^t\beta \cdot Z \in Z \cdot Sp(d/2;R)\}.$

Further, by restricting the extension given in Lemma 10.3 , we see that

(7.6b) every $[\pi_{0,Z,\gamma}] \in I_{d;c}(R)^{\wedge}$ extends to a class $[\tilde{\pi}_{0,Z,\gamma}] \in \{I_{d;c}(R) \cdot M_Z\}^{\wedge}.$

Now, from the little-group method,

   7.7.   Theorem. $J_{d;e,a,b}(R)^{\wedge}$ *is the disjoint union of non-empty subsets*
$S_J(\ell,\{A\})$ *where* ℓ *and* m *are non-negative integers such that*

$$\ell + 2m \leqslant e + a + b \quad \underline{and} \quad 2(\ell + m) \leqslant d$$

and where $\{A\}$ *is an* $L_{e;a,b}(R)$*-equivalence (under* $A \to {}^t\beta^{-1} \cdot A \cdot \beta^{-1}$*) class of*
$(e + a + b) \times (e + a + b)$ *antisymmetric real matrices of rank* $2m$*.*
$S_J(\ell,\{A\})$ *is parameterized by*

$$G_{\ell;d+2(e+a+b)-2(\ell+m)}(R)^{\wedge} \times \hat{M}_Z$$

under

$$([\gamma],[\mu]) \leftrightarrow [Ind_{I_{d;e+a+b}(R) \cdot M_Z} \uparrow J_{d;e,a,b}(R) (\tilde{\pi}_{0,Z,\gamma} \otimes \mu)]$$

where $Z \in R^{(e+a+b) \times d}$ *with* $\beta(Z.Z) = A$ *and* rank $Z = \ell + 2m$*.*

    This reduces the determination of $J_{d;e,a,b}(R)^{\wedge}$ to the determination
of the $\hat{M}_Z$. The M_Z are groups of similar type, but with smaller matrices,
as we will see in a moment. Thus, by recursion on the size of the matrices
we may suppose the $\hat{M}_Z$ known.

To complete the discussion for $J_{d;e,a,b}(R)$, we set $c = e + a + b$ and note that there exist

(7.8a)   a symmetric bilinear form β on R^c of "signature" (a,b,e)

and

(7.8b)   an antisymmetric bilinear form α on R^c of rank $2m$

such that

(7.8c) $M_Z = \{g \in GL(c;R): \ g \ \text{preserves both} \ \alpha \ \text{and} \ \beta\}.$

In effect, $L_{e;a,b}(R) = \{g \in GL(c;R): \ g \ \text{preserves} \ \beta\}$, and condition (7.6a) that ${}^t g \cdot Z \in Z \cdot Sp(d/2;R)$ simply says that g preserves the anti-symmetric form α with matrix $\beta(Z,Z)$. Now split $R^c = U \oplus V \oplus W$ where

$U \oplus V$ is the null space of α, so $\alpha|_{W \times W}$ is nondegenerate,

U is the null space of $\beta|_{(U+V) \times (U+V)}$, so $\beta|_{V \times V}$ is nondegenerate,

W is orthogonal to V relative to β.

If $g \in M_Z$ now it preserves U and $U \oplus V$, and $g(W) \subset U \oplus W$ because $\beta(V,W) = 0$ and β is nondegenerate on V. Thus an element $g \in M_Z$ is given schematically by

$$
\begin{bmatrix} g_{11} & g_{12} & g_{13} \\ 0 & g_{22} & 0 \\ 0 & 0 & g_{33} \end{bmatrix} \in \begin{bmatrix} GL(U) & U \otimes V^* & U \otimes W^* \\ 0 & O(\beta|_{V \times V}) & 0 \\ 0 & 0 & Sp(\alpha|_{W \times W}) \end{bmatrix}
$$

where g_{13} and g_{33} are related by $\beta(gw, gw') = \beta(w, w')$ for all w, $w' \in W$. Set

$$n_1 = \dim U, \quad n_2 = \dim V, \quad n_3 = \dim W \quad \text{and} \quad (m_1, m_2) = \text{signature } (\beta|_{V \times V}).$$

Then

$$M_Z \cong I'_{n_1, n_2; n_3}(\mathbb{R}) \cdot L_{n_1; m_1, m_2}(\mathbb{R})$$

where $L_{n_1; m_1, m_2}(\mathbb{R}) = \{(\chi; \sigma, \tau) \in L_{n_1, n_2}(\mathbb{R}): \tau \in O(m_1, m_2)\}$ as in (6.9d), and where $I'_{n_1, n_2; n_3}(\mathbb{R})$ is the subgroup $(g|_{U+V} = \text{identity})$ of M_Z, which sits naturally in $\mathbb{R}^{n_1 \times n_3} \cdot Sp(\frac{1}{2}n_3; \mathbb{R})$ under

$$\begin{bmatrix} I & 0 & g_{13} \\ 0 & I & 0 \\ 0 & 0 & g_{33} \end{bmatrix} \rightarrow \begin{bmatrix} I & g_{13} \\ 0 & g_{33} \end{bmatrix} \rightarrow (g_{13} \, g_{33}^{-1}, \, g_{33}).$$

Given specific α and β, we now have enough information to determine $\hat{M}_Z$.

Now $J_{d;e,a,b}(\mathbb{R})^{\wedge}$ is determined.

§7B. The case $F = \mathbb{C}$.

Looking back at (2.8) and Corollary 2.14, we see that

$$(7.9a) \qquad I_{d', d''; c}(\mathbb{C}) \cong G_{c; d', d''}(\mathbb{C}) / \text{Im } \mathbb{C}^{c \times c}$$

where, for a totally isotropic c-dimensional subspace $E \subset \mathbb{C}^{c+d', c+d''}$

(7.9b) $G_{c;d',d''}(C) = \{g \in U(c+d',c+d''): g|_E = \text{identity}\}.$

Further, $\text{Im } C^{c \times c}$ is central in the unipotent radical $N_{c;d',d''}(C)$ of $G_{c;d',d''}(C)$, and $N_{c;d',d''}(C)/\text{Im } C^{c \times c}$ is the vector group $C^{c \times (d',d'')}$. In the notation (5.9), that tells us

(7.10) $I_{d',d'';c}(C)^\wedge = \{[\pi_{D,Z,\gamma}] \in G_{c;d',d''}(C)^\wedge : D = 0\}.$

Now Theorem 5.12 specializes to

7.11. Proposition. $I_{d',d'';c}(C)^\wedge$ _is the disjoint union of non-empty subsets_ $S_I(\ell,m,n)$ _where_ ℓ, m _and_ n _are non-negative integers such that_

(*) $\ell + m + n \leq c, \quad \ell + n \leq d' \quad and \quad m + n \leq d'',$

given as follows. Let $\mathcal{H}_{\ell,m}$ _denote the space of all_ $c \times c$ _complex hermitian matrices of "signature"_ $(\ell,m,c-\ell-m)$. _Then_ $S_I(\ell,m,n)$ _is parameterized by_

$$\mathcal{H}_{\ell,m} \times G_{n;d'-\ell-n,d''-m-n}(C)^\wedge$$

under

$$(H,[\gamma]) \leftrightarrow [\pi_{0,Z,\gamma}] \quad in\ the\ notation\ (7.10)\ and\ (5.9).$$

where $Z \in C^{c \times (d',d'')}$ _with_ $\mathcal{H}(Z,Z) = H$ _and_ $\text{rank } Z = \ell + m + n.$

The $L_{e;a,b}(C)$-stabilizer of $[\pi_{0,Z,\gamma}] \in I_{d',d'';e+a+b}(C)^\wedge$ is obtained by specializing Lemma 6.1 to the case $D = 0$. Let $c = e+a+b$ and

$z \in C^{c\times(d',d'')}$ with rank $Z = \ell + m + n$ and $\mathcal{H}(Z,Z)$ of "signature"

$(\ell,m,c-\ell-m)$, as in Proposition 7.11; then that stabilizer is

(7.12a) $\qquad M_Z = \{\beta \in L_{e;a,b}(C) \subset GL(c;C): \beta^* Z \in Z \cdot U(d',d'')\}.$

Restricting the extension given by Lemma 6.6, we see that

(7.12b) every $[\pi_{0,Z,\gamma}] \in I_{d',d'';c}(C)^\wedge$ extends to a class

$$[\tilde{\pi}_{0,Z,\gamma}] \in \{I_{d',d'';c}(C) \cdot M_Z\}^\wedge.$$

Thus, by Mackey little-group method,

$\qquad$ 7.13. **Theorem.** $J_{d',d'';e,a,b}(C)^\wedge$ _is the disjoint union of non-empty_
subsets $S_J(n;\{H\})$ _where_ ℓ, m _and_ n _are non-negative integers such_
that

$$\ell + m + n \leqslant e + a + b, \quad \ell + n \leqslant d' \;\underline{and}\; m + n \leqslant d''$$

and where $\{H\}$ _is an_ $L_{e;a,b}(C)$-_equivalence_ _(under_ $(H \to (\beta^*)^{-1}H\beta^{-1})$ _class_
_of_ $(e+a+b) \times (e+a+b)$ _complex hermitian matrices of "signature"_
$(\ell,m,e+a+b-\ell-m)$. $S_J(n;\{H\})$ _is parameterized by_

$$G_{n;d'-\ell-n,d''-m-n}(C)^\wedge \times \hat{M}_Z$$

under

$$([\gamma],[\mu]) \leftrightarrow [\mathrm{Ind}_{I_{d',d'';e+a+b}(C)\cdot M_Z \uparrow J_{d',d'';e,a,b}(C)}(\tilde{\pi}_{0,Z,\gamma} \otimes \mu)]$$

where $Z \in C^{(e+a+b)\times(d',d'')}$ _with_ $\mathcal{H}(Z,Z) = H$ _and_ rank $Z = \ell + m + n$.

This reduces the determination of $J_{d',d'';e,a,b}(\mathbb{C})\hat{\ }$ to the determination of $\hat{M}_Z$. In a moment we will see that the M_Z are groups of similar type, but with smaller matrices. Thus by recursion of the size of the matrices we may suppose the $\hat{M}_Z$ known.

To complete this discussion for $J_{d',d'';e,a,b}(\mathbb{C})\hat{\ }$, we set $c = e+a+b$ as before and note that we have

(7.14a)   an hermitian form k on $\mathbb{C}^c$ of "signature" (a,b,e)

and

(7.14b)   another hermitian form h on $\mathbb{C}^c$ of "signature" $(\ell,m,c-\ell-m)$

such that

(7.14c) $M_Z = \{g \in GL(c;\mathbb{C}): g \text{ preserves both } h \text{ and } k\}$.

For $L_{e;a,b}(\mathbb{C}) = \{g \in GL(c;\mathbb{C}): g \text{ preserves } k\}$, and condition (7.12a) that $g^*Z \in Z \cdot U(d',d'')$ says that g preserves the hermitian form h with matrix $\mathcal{H}(Z,Z)$. Now split $\mathbb{C}^c = U \oplus V \oplus W$ as before, where

$U \oplus V$ is the null space of h, so $h|_{W \times W}$ is nondegenerate,

U is the null space of $k|_{(U+V) \times (U+V)}$, so $k|_{V \times V}$ is nondegenerate,

W is orthogonal to V relative to k.

Every $g \in M_Z$ preserves U and $U \oplus V$ and maps W into $U \oplus W$, so it is given schematically by

$$\begin{bmatrix} g_{11} & g_{12} & g_{13} \\ 0 & g_{22} & 0 \\ 0 & 0 & g_{33} \end{bmatrix} \in \begin{bmatrix} GL(U) & U \otimes V^* & U \otimes W^* \\ 0 & U(k|_{V \times V}) & 0 \\ 0 & 0 & U(h|_{W \times W}) \end{bmatrix}$$

with g_{13} and g_{33} related by: $k(gw, gw') = k(w, w')$ for all $w, w' \in W$. Let

$$n_1 = \dim U, \quad n_2 = \dim V, \quad n_3 = \dim W \quad \text{and} \quad (m_1, m_2) = \text{signature} (k|_{V \times V}).$$

Then $M_Z \cong I'_{n_1, n_2; n_3}(R) \cdot L_{n_1; m_1, m_2}(C)$ where

$L_{n_1; m_1, m_2}(C) = \{(X; \sigma, \tau) \in L_{n_1, n_2}(R) : \tau \in U(m_1, m_2)\}$ as in (6.9d), and where

$I'_{n_1, n_2; n_3}(C)$ is the subgroup $(g|_{U+V} = \text{identity})$ of M_Z, naturally

isomorphic to a subgroup of $C^{n_1 \times n_3} \cdot U(\ell, m)$ under

$$\begin{bmatrix} I & 0 & g_{13} \\ 0 & I & 0 \\ 0 & 0 & g_{33} \end{bmatrix} \rightarrow \begin{bmatrix} I & g_{13} \\ 0 & g_{33} \end{bmatrix} \rightarrow (g_{13} g_{33}^{-1}, g_{33}).$$

Given specific h and k, we now have enough to determine $\hat{M}_Z$, and so $J_{d', d''; e, a, b}(C)^{\wedge}$ is determined.

§7C. The Case $F = Q$.

Looking ahead to (13.7) and Theorem 13.9, we are going to see that

$$(7.15a) \qquad \qquad I_{d; c}(Q) \cong G^*_{2c, d} / \mathcal{S}\mathcal{O}^*(2c)$$

where, for a doubly totally isotropic $2c$-dimensional subspace $E \subset C^{2d + 4c}$

(7.15b) $G^*_{2c,d} = \{g \in SO^*(2d + 4c): g|_E = \text{identity}\}.$

For that, we consider the map $Q^{d \times c} \to C^{2c \times d}$ given by $A + Bj \to {}^t(A,B)$ where $A,B \in C^{d \times c}$, and we note from the discussion surrounding (14.19) that this induces the correct map from $SO^*(2d) \subset GL(d;Q)$ to $SO^*(2d) \subset GL(2d;C)$.

In (7.15), $\mathfrak{so}^*(2c)$ is central in the unipotent radical $N^*_{2c;d}$ of $G^*_{2c,d}$, the quotient $N^*_{2c,d}/\mathfrak{so}^*(2c)$ being the real vector group $Q^{d \times c} \cong C^{2c \times d}$. In the notation (14.22), that gives us

(7.16) $I_{d;c}(Q)^\wedge = \{[\pi_{D,Z,\gamma}] \in (G^*_{2c;d})^\wedge : D = 0\}.$

Now Theorem 14.24 specializes to

**7.17. Proposition.** $I_{d;c}(Q)^\wedge$ is the disjoint union of non-empty subsets $S_I(\ell,m)$ where ℓ and m are non-negative integers such that

(*) $\ell + m \leqslant c$ _and_ $2\ell + m \leqslant d,$

given as follows. Let $\mathcal{M}_m$ denote the space of all $2c \times 2c$ complex matrices M of rank $2m$ such that

$$\begin{bmatrix} 0 & I \\ -I & 0 \end{bmatrix} \cdot \bar{M} \cdot \begin{bmatrix} 0 & I \\ -I & 0 \end{bmatrix}^{-1} = M \quad \underline{and} \quad M \cdot \begin{bmatrix} 0 & I \\ I & 0 \end{bmatrix} = \begin{bmatrix} 0 & I \\ I & 0 \end{bmatrix} \cdot M^* .$$

Then $S_I(\ell,m)$ is parameterized by $\mathcal{M}_m \times (G^*_{2\ell;d-m-2\ell})^\wedge$ under

$(M,[\gamma]) \leftrightarrow [\pi_{0,Z,\gamma}]$ _in the notation_ (14.22) _and_ (7.16)

where $Z \in C^{2c \times d}$ with $\mathcal{M}(Z,Z) = M$ and $\frac{1}{2}$ rank $\left(Z, \begin{bmatrix} 0 & I \\ -I & 0 \end{bmatrix} \bar{Z}\right) = \ell + m.$

The $L_{e;a,b}(Q)$-stabilizer of $[\pi_{0,Z,\gamma}] \in I_{d;e+a+b}(Q)^{\wedge}$, $c = e + a + b$, is obtained by specializing Lemma 15.1 to the case $D = 0$. That stabilizer is

$$(7.18a) \quad M_Z = \{\beta \in L_{e;a,b}(Q) \subset GL(c;Q): \beta'Z = Z_g \text{ for some } g \in SO^*(2d)\}.$$

The extension given in Lemma 15.3 restricts to give us

$$(7.18b) \quad \text{every } [\pi_{0,Z,\gamma}] \in I_{d;c}(Q)^{\wedge} \text{ extends to some } [\tilde{\pi}_{0,Z,\gamma}] \in \{I_{d;c}(Q) \cdot M_Z\}^{\wedge}.$$

Now the little-group method tells us

7.19. Theorem. $J_{d;e,a,b}(Q)^{\wedge}$ *is the disjoint union of non-empty subsets* $S_J(\ell;\{M\})$, *where* ℓ *and* m *are non-negative integers such that*

$$\ell + m \leqslant e + a + b \quad \underline{and} \quad 2\ell + m \leqslant d ,$$

and where $\{M\}$ *is an* $L_{e;a,b}(Q)$-*equivalence class of* $2(e+a+b) \times 2(e+a+b)$ *complex matrices* M *of rank* $2m$ *such that*

$$\begin{bmatrix} 0 & I \\ -I & 0 \end{bmatrix} \cdot \bar{M} \cdot \begin{bmatrix} 0 & I \\ -I & 0 \end{bmatrix}^{-1} = M \quad \underline{and} \quad M \cdot \begin{bmatrix} 0 & I \\ I & 0 \end{bmatrix} = \begin{bmatrix} 0 & I \\ I & 0 \end{bmatrix} \cdot M^*.$$

$S_J(\ell;\{M\})$ *is parameterized by* $(G^*_{2\ell;d-m-2\ell})^{\wedge} \times \hat{M}_Z$ *under*

$$([\gamma],[\mu]) \leftrightarrow [\text{Ind}_{I_{d;e+a+b}(Q) \cdot M_Z \uparrow J_{d;e,a,b}(Q)}(\tilde{\pi}_{0,Z,\gamma} \otimes \mu)]$$

where $Z \in C^{2(e+a+b) \times d}$ *with* $\mathcal{M}(Z,Z) = M$ *and* $\frac{1}{2} \text{rank}\left(Z, \begin{pmatrix} 0 & I \\ -I & 0 \end{pmatrix}\bar{Z}\right) = \ell + m.$

The determination of $J_{d;e,a,b}(Q)^{\wedge}$ now is reduced to the determination of the $\hat{M}_Z$. The M_Z are groups of similar type constructed with smaller matrices, as we will see in a moment, so we may suppose $\hat{M}_Z$ known by recursion on the size of the matrices.

To complete the discussion of $J_{d;e,a,b}(Q)$, set $c = e + a + b$ and note that we have

(7.20a)   a hermitian form β on Q^c of "signature" (a,b,e)

and

(7.20b)   a skew hermitian form φ on Q^c

such that

(7.20c) $M_Z = \{g \in GL(c;Q): \; g \text{ preserves both } \beta \text{ and } \varphi\}.$

For this, first observe $L_{e;a,b}(Q) = \{g \in GL(c;Q): \; g \text{ preserves } \beta\}$. Now the condition (7.18) that $g'Z = Z_h$ for some element $h \in SO^*(2d)$ just says $g'\left(Z, \begin{pmatrix} 0 & I \\ -I & 0 \end{pmatrix} \bar{Z}\right) \in \left(Z, \begin{pmatrix} 0 & I \\ -I & 0 \end{pmatrix} \bar{Z}\right) \cdot SO^*(2d)$. In other words, it says that g preserves the skew hermitian form on Q^c with matrix $(\kappa(z_r, z_s))$, where κ is the skew hermitian form defining $SO^*(2c)$ in the proof of Lemma 6.1, and where the z_r are columns of the quaternionic matrix corresponding to ${}^t\left(Z, \begin{bmatrix} 0 & I \\ -I & 0 \end{bmatrix} \bar{Z}\right)$. As before, we split $Q^c = U \oplus V \oplus W$ where

$U \oplus V$ is the null space of φ, so $\varphi|_{W \times W}$ is nondegenerate,
   U is the null space of $\beta|_{(U+V) \times (U+V)}$, so $\beta|_{V \times V}$ is nondegenerate
         W is orthogonal to V relative to β.

Every element of M_Z preserves U and $U \oplus V$ and maps W into $U \oplus W$, so the general $g \in M_Z$ is given schematically by

$$\begin{bmatrix} g_{11} & g_{12} & g_{13} \\ 0 & g_{22} & 0 \\ 0 & 0 & g_{33} \end{bmatrix} \in \begin{bmatrix} GL(U) & U \otimes V^* & U \otimes W^* \\ 0 & Sp(\beta|_{V\times V}) & 0 \\ 0 & 0 & SO^*(\varphi|_{W\times W}) \end{bmatrix}$$

subject to the condition on g_{13} and g_{33} that $\beta(gw,gw') = \beta(w,w')$ for all $w, w' \in W$. We set

$$n_1 = \dim U, \quad n_2 = \dim V, \quad n_3 = \dim W \quad \text{and} \quad (m_1,m_2) = \text{signature } (\beta|_{V\times V}).$$

Then $M_Z \cong I'_{n_1,n_2;n_2}(Q) \cdot L_{n_1;m_1,m_2}(Q)$ where

$L_{n_1;m_1,m_2}(Q)$: $\{(\chi;\sigma,\tau) \in L_{n_1;n_2}(Q): \tau \in Sp(m_1,m_2)\}$ as in (6.9d), and where

$I'_{n_1,n_2;n_3}(Q)$ is the subgroup $(g|_{U+V} = \text{identity})$ which sits naturally in $Q^{n_1 \times n_3} \cdot SO^*(\frac{1}{2}n_3)$ under

$$\begin{bmatrix} I & 0 & g_{13} \\ 0 & I & 0 \\ 0 & 0 & g_{33} \end{bmatrix} \rightarrow \begin{bmatrix} I & g_{13} \\ 0 & g_{33} \end{bmatrix} \rightarrow (g_{13}g_{33}^{-1}, g_{33}).$$

Given specific β and φ, this suffices to determine $\hat{M}_Z$. Finally, now, $J_{d;e,a,b}(Q)^{\wedge}$ is determined.

Part III. Symplectic Groups.

§8. Parabolic Subgroups of Symplectic and Metaplectic Groups

Let F be a commutative field, eventually R or C. Any nondegenerate antisymmetric bilinear form on F^{2n} is equivalent to

$$(8.1a) \qquad \alpha(x,y) = \sum_{1 \leqslant \ell \leqslant n} (x^\ell y^{n+\ell} - x^{n+\ell} y^\ell).$$

This defines the *symplectic group*

$$(8.1b) \qquad Sp(n;F) = \{g \in GL(2n;F): \alpha(gx,gy) = \alpha(x,y), \quad \text{all} \quad x,y\}.$$

$Sp(n;C)$ is simply connected and we set $Mp(n;C) = Sp(n;C)$. $Sp(n;R)$ has a unique 2-sheeted covering group, which we denote $Mp(n;R)$. Thus, for $F = R$ or C we have

$$(8.1c) \qquad \varepsilon: Mp(n;F) \to Sp(n;F), \quad \text{covering by } \textit{metaplectic group}.$$

For general $Mp(n;F)$, see the remarks following (8.5) below.

A linear subspace $E \subset F^{2n}$ is called *totally isotropic* if $\alpha(E,E) = 0$.
The parabolic subgroups of $Sp(n;F)$ are the

$$(8.2a) \qquad P_{E_1,\ldots,E_k} = \{g \in Sp(n;F): gE_\ell = E_\ell \quad \text{for} \quad 1 \leqslant \ell \leqslant k\}$$

where $0 \neq E_1 \subsetneq \ldots \subsetneq E_k$ is an increasing sequence of totally isotropic subspaces. The sequence $\{\dim E_\ell\}$ determines the $Sp(n;F)$-equivalence class of the sequence $\{E_\ell\}$, hence also the $Sp(n;F)$-conjugacy class of $P_{E_1,\ldots,E_k}$. In particular, the maximal parabolic subgroups of $Sp(n;F)$ are the

(8.2b) $P_E = \{g \in Sp(n;F): gE = E\}$, $E \subset F^{2n}$ totally isotropic,

and there are just n conjugacy classes as dim E = 1,2,...,n.

The parabolic subgroups of the metaplectic groups are the

(8.3a) $\tilde{P}_{E_1,\ldots,E_k} = \varepsilon^{-1} \cdot P_{E_1,\ldots,E_k}$.

In particular, the maximal parabolic subgroups are the

(8.3b) $\tilde{P}_E = \varepsilon^{-1} \cdot P_E$, E nonzero totally isotropic in F^{2n}.

We are going to work out the structure of the maximal parabolic sub-
groups $P_E \subset Sp(n;F)$ and $\tilde{P}_E \subset Mp(n;F)$, proceeding as in §§2 and 3. The
result is true in considerably greater generality, but we minimize techni-
calities by writing down a Lie algebra proof valid only in characteristic
zero, for our principal concern is the cases F = R and F = C. The reader
acquainted with linear algebraic groups can easily make the extension.

Let s and t be non-negative integers. Sym $F^{s \times s}$ denotes the space of
s × s symmetric matrices over F. Now

(8.4a) $\mathscr{A}((X,Y), (X',Y')) = \frac{1}{2}\{X \cdot {}^tY' + Y' \cdot {}^tX - X' \cdot {}^tY - Y \cdot {}^tX'\}$

defines an antisymmetric bilinear map $\mathscr{A}: F^{s \times 2t} \times F^{s \times 2t} \to Sym\ F^{s \times s}$, which
in turn defines a group and its Lie algebra:

(8.4b) $N_{s;2t}(F) = Sym\ F^{s \times s} + F^{s \times 2t}$ with

         $(D,Z)(D',Z') = (D + D' + \mathscr{A}(Z,Z'),\ Z + Z')$,

(8.4c) $n_{s;2t}(F) = Sym\ F^{s \times s} + F^{s \times 2t}$ with $[(D,Z),(D',Z')] = (2\mathscr{A}(Z,Z'),0)$.

$N_{1;2t}(F)$ is the usual Heisenberg group over F. We have semidirect product groups

$$(8.5a) \qquad G_{s;2t}(F) = N_{s;2t}(F) \cdot Sp(t;F) \quad \text{and} \quad \tilde{G}_{s;2t}(F) = N_{s;2t}(F) \cdot Mp(t;F)$$

with group law $(D,Z;g)(D',Z';g') = (D + D' + \mathcal{A}(Z,Z'g^{-1}), \ Z + Z'g^{-1};gg')$.
In the metaplectic case, $Z'g^{-1}$ means $Z' \cdot \varepsilon(g^{-1})$. We also have

$$(8.5b) \qquad P_{s;2t}(F) = G_{s;2t}(F) \cdot GL(s;F) = N_{s;2t}(F) \cdot \{Sp(t;F) \times GL(s;F)\}$$

and

$$(8.5c) \qquad \tilde{P}_{s;2t}(F) = \tilde{G}_{s;2t}(F) \cdot GL(s;F) = N_{s;2t}(F) \cdot \{Mp(t;F) \times GL(s;F)\}$$

with group law

$$(D,Z;g,\gamma)(D',Z';g',\gamma') = (D + \gamma D' \cdot {}^t\gamma + \mathcal{A}(Z,\gamma Z'g^{-1}), \ Z + \gamma Z'g^{-1}; \ gg',\gamma\gamma').$$

$Sp(n;F)$ is the group of all F-automorphisms of $N_{1;2n}(F)$ that act trivially on the center. Following an idea of Weil [24], whenever F is a locally compact field of characteristic $\neq 2$ one can define a metaplectic group $Mp(n;F)$ as the minimal central extension of $Sp(n;F)$ such that

> if $[\pi] \in N_{1;2n}(F)^{\wedge}$ has nontrivial central character, then
> $[\pi]$ extends to a class $[\tilde{\pi}] \in \{N_{1;2n}(F) \cdot Mp(n;F)\}^{\wedge}$ on the same space.

It is trivial for $F = \mathbb{C}$, and due to Shale [21] for $F = R$, that this is consistent with our definitions above. That consistency is the key point in Theorem 9.19 of the next section, where metaplectic groups are forced on us because representations of $N_{s;2t}(R)$ often do not extend to their

$G_{s;2t}(R)$-stabilizers. With these comments in mind, the interested reader will see how our results go over to more general situations.

Here is the structure theorem for the maximal parabolic subgroups of symplectic and metaplectic groups.

8.6. Theorem. *Let* E *be a totally isotropic subspace of dimension* s *in* F^{2n}. *Let* $P_E \subset Sp(n;F)$ *and* $\widetilde{P}_E \subset Mp(n;F)$ *be the corresponding maximal parabolic subgroups. Then there are Lie group/linear algebraic group isomorphisms*

$$(8.7a) \qquad \varphi : P_{s;2(n-s)}(F) \to P_E \quad \underline{and} \quad \widetilde{\varphi} : \widetilde{P}_{s;2(n-s)}(F) \to \widetilde{P}_E$$

that carry $N_{s;2(n-s)}(F)$ *to the unipotent radical and* $Sp(n-s;F) \times GL(s;F)$, *or* $Mp(n-s;F) \times GL(s;F)$, *to a reductive complement.*

The Langlands decompositions $P_E = MAN$ *and* $\widetilde{P}_E = \widetilde{M}AN$ *are reflected by*

$$(8.7b) \quad \begin{cases} M = \varphi \cdot \{Sp(n-s;F) \times GL'(s;F)\} \quad \underline{and} \quad \widetilde{M} = \widetilde{\varphi} \cdot \{Mp(n-s;F) \times GL'(s;F)\}, \\[2mm] N = \varphi \cdot N_{s;2(n-s)}(F) = \widetilde{\varphi} \cdot N_{s;2(n-s)}(F), \quad \underline{and} \quad A = \varphi \cdot R^+ = \widetilde{\varphi} \cdot R^+. \end{cases}$$

Further, φ *and* $\widetilde{\varphi}$ *have restrictions*

$$\varphi : G_{s;2(n-s)}(F) \to \{g \in Sp(n;F): g|_E = \text{identity}\} \quad \underline{and}$$

$$\widetilde{\varphi} : \widetilde{G}_{s;2(n-s)}(F) \to \{\widetilde{g} \in Mp(n;F): \varepsilon(\widetilde{g})|_E = \text{identity}\} \quad .$$

We note some immediate consequences.

8.8. Corollary. *Maximal parabolic subgroups* $P_E \subset Sp(n;F)$ *and* $\widetilde{P}_E \subset Mp(n;F)$ *have abelian unipotent radical if and only if* $\dim E = n$.

8.9. Corollary. *Maximal parabolic subgroups* $P_E \subset Sp(n;C) = Mp(n;C)$
are cuspidal just when $n = \dim E = 1$.

8.10. Corollary. *Maximal parabolic subgroups* $P_E \subset Sp(n;R)$
$\widetilde{P}_E \subset Mp(n;R)$ *are cuspidal just when* E *has dimension* 1 *or* 2.

We now turn to the proof of Theorem 8.6. We only write out the proof
for $Sp(n;F)$; at every step the corresponding fact goes over to $Mp(n;F)$.
Let $\{e_1,\ldots,e_{2n}\}$ denote the standard basis in which α is given by (8.1a).

We may suppose that $\{e_1,\ldots,e_s\}$ is a basis of E. Denote

(8.11a) $E' = span \{e_{n+1},\ldots,e_{n+s}\}$,

(8.11b) $V = E + E' = span \{e_1,\ldots,e_s; e_{n+1},\ldots,e_{n+s}\}$,

(8.11c) $W = V^{\perp} = span \{e_{s+1},\ldots,e_n; e_{n+s+1},\ldots,e_{2n}\}$.

Then $F^{2n} = V \oplus W$, orthogonal under α. Also denote

(8.12a) $\mathfrak{g}$: Lie algebra of $Sp(n;F)$,

(8.12b) $\mathfrak{p}$: Lie algebra of $P_E = \{g \in Sp(n;F): gE = E\}$,

(8.12c) $\mathfrak{l}$: Lie algebra of $L_E = \{g \in Sp(n;F): g|_E = identity\}$.

Then

(8.13a) $\mathfrak{g} = \{\xi: F^{2n} \to F^{2n} \text{ linear}: \alpha(\xi x,y) + \alpha(x,\xi y) = 0, \text{ all } x,y\}$,

(8.13b) $\mathfrak{p} = \{\xi \in \mathfrak{g}: \xi E \subset E\}$, and $\mathfrak{l} = \{\xi \in \mathfrak{g}: \xi E = 0\}$.

8.14. Lemma. $\mathfrak{p}$ *is direct sum of its linear subspaces*

$$\mathfrak{p}_r^W = \{\xi \in \mathfrak{g}: \xi V = 0 \text{ } \underline{and} \text{ } \xi W \subset W\},$$

$$\mathfrak{p}_r^V = \{\xi \in \mathfrak{g}: \xi E \subset E, \xi E' \subset E' \text{ } \underline{and} \text{ } \xi W = 0\} \text{ },$$

$$\mathfrak{p}_n^1 = \{\xi \in \mathfrak{g}: \xi E' \subset W, \xi W \subset E \text{ } \underline{and} \text{ } \xi E = 0\} \text{ },$$

$$\mathfrak{p}_n^2 = \{\xi \in \mathfrak{g}: \xi E' \subset E, \xi E = 0 \text{ } \underline{and} \text{ } \xi W = 0\} \text{ };$$

and $\mathfrak{l} = \mathfrak{p}_r^W + \mathfrak{p}_n^1 + \mathfrak{p}_n^2.$

Proof: Replace h by α in the proof of Lemma 3.4.

$$q.e.d.$$

Apply (8.13a) to the Lie algebra $\mathfrak{sp}(n-s;F)$ of the symplectic group $Sp(n-s;F)$ of W. It says

$$(8.15) \qquad \mathfrak{sp}(n-s;F) = \left\{ \begin{bmatrix} a & b \\ c & -\,^t a \end{bmatrix} \ b = \,^t b \ \text{and} \ c = \,^t c \right\}$$

where $a, b, c \in F^{(n-s)\times(n-s)}.$

8.16. Lemma. _Define_ $\psi_r \colon \mathfrak{sp}(n-s;F) \oplus \mathfrak{gl}(s;F) \to \mathfrak{gl}(2n;F)$ _by_

$$\psi_r\!\left(\begin{bmatrix} a & b \\ c & -\,^t a \end{bmatrix}, \ \gamma \right) \ = \ \begin{bmatrix} \gamma & 0 & 0 & 0 \\ 0 & a & 0 & b \\ 0 & 0 & -\,^t\gamma & 0 \\ 0 & c & 0 & -\,^t a \end{bmatrix}.$$

Then $\psi_r \cdot \mathfrak{sp}(n-s;F) = \mathfrak{p}_r^W, \ \psi_r \cdot \mathfrak{gl}(s;F) = \mathfrak{p}_r^V$ _and_ ψ_r _is a Lie algebra isomorphism onto_ $\mathfrak{p}_r = \mathfrak{p}_r^W \oplus \mathfrak{p}_r^V.$

Proof: Let $\xi \in \mathfrak{p}_r.$ Then $\xi W \subset W$ with $\xi|_W$ in the Lie algebra of the symplectic group there, say $\xi|_W$ has matrix $\begin{bmatrix} a & b \\ c & -\,^t a \end{bmatrix} \in \mathfrak{sp}(n-s;F).$ Similarly, $\xi|_V$ has matrix $\begin{bmatrix} \gamma & \delta \\ \varepsilon & -\,^t\gamma \end{bmatrix} \in \mathfrak{sp}(s;F),$ $\xi E \subset E$ says $\varepsilon = 0,$ and $\varepsilon E' \subset E'$ says $\delta = 0.$ Thus ξ is the image of $\psi_r.$ As ψ_r clearly is an injective homomorphism, the Lemma is proved.

$$q.e.d.$$

8.17. Lemma. _Define_ $\psi_n \colon \mathfrak{n}_{s;2(n-s)}(F) \to \mathfrak{gl}(2n;F)$ _by_

$$\psi_n(D,(A,B)) = \begin{bmatrix} 0 & A & D & B \\ 0 & 0 & {}^tB & 0 \\ 0 & 0 & 0 & 0 \\ 0 & 0 & -{}^tA & 0 \end{bmatrix}$$

Then $\psi_n \cdot \text{Sym } F^{s \times s} = \mathfrak{p}_n^2$, $\psi_n \cdot F^{s \times 2(n-s)} = \mathfrak{p}_n^1$, _and_ ψ_n _is a Lie algebra isomorphism onto_ $\mathfrak{p}_n = \mathfrak{p}_n^2 + \mathfrak{p}_n^1$. _Further, the image of_ ψ_n _consists of nilpotent matrices._

Proof. Denote $\zeta_D = \psi_n(D,0)$ and $\eta_{A,B} = \psi_n(0,(A,B))$. By direct calculation,

$$\zeta_D \zeta_{D'} = 0, \quad \text{in particular} \quad \zeta_D^2 = 0 \quad \text{and} \quad [\zeta_D, \zeta_{D'}] = 0.$$

Again by direct calculation,

$$\zeta_D \eta_{A,B} = 0 = \eta_{A,B} \zeta_D, \quad \text{in particular} \quad [\zeta_D, \eta_{A,B}] = 0.$$

Finally by direct calculation with $Z = (A,B)$ and $Z' = (A',B')$,

$$\eta_{A,B} \eta_{A',B'} = \begin{bmatrix} 0 & 0 & A{}^tB' - B{}^tA' & 0 \\ 0 & 0 & 0 & 0 \\ 0 & 0 & 0 & 0 \\ 0 & 0 & 0 & 0 \end{bmatrix}.$$

In particular,

$$\eta_{A,B}^3 = 0 \quad \text{and} \quad [\eta_Z, \eta_{Z'}] = \zeta_{2\mathscr{A}(Z,Z')}.$$

We conclude that ψ_n is a Lie algebra isomorphism onto its image and that the image consists of nilpotent matrices.

Let $\zeta \in \mathfrak{p}_n^2$. Then $\zeta \in \mathfrak{F}$, $\zeta E' \subset E$ and $\zeta(E + W) = 0$ force $\zeta = \zeta_D$ for some $D \in \text{Sym } F^{s \times s}$.

Let $\eta \in \mathfrak{p}_n^1$. Then $\eta \in \mathfrak{s}$, $\eta E' \subset W$, $\eta W \subset E$ and $\eta E = 0$ force $\eta = \eta_{A,B}$ for some $(A,B) \in F^{s \times 2(n-s)}$.

Now ψ_n has image $\mathfrak{p}_n$ as required.   _q.e.d._

Lemmas 8.16 and 8.17 give a vector space isomorphism $\psi \colon \mathfrak{p}_{s;2(n-s)}(F) \to \mathfrak{p}$ such that $\psi \cdot \mathfrak{g}_{s;2(n-s)}(F) = \mathfrak{l}$ and the restrictions

$$\psi_n \colon \mathfrak{n}_{s;2(n-s)}(F) \to \mathfrak{p}_n \quad \text{and} \quad \psi_r \colon \mathfrak{sp}(n-s;F) \oplus \mathfrak{gl}(s;F) \to \mathfrak{p}_r$$

are Lie algebra isomorphisms. Direct calculation gives

8.18. <u>Lemma</u>. _Let_ $\begin{bmatrix} a & b \\ c & -{}^t a \end{bmatrix} \in \mathfrak{sp}(n-s;F)$, $\gamma \in \mathfrak{gl}(s;F)$, $D \in \mathrm{Sym}\ F^{s \times s}$ _and_ $\zeta_D = \psi_n(D,0)$, _and_ $Z = (A,B) \in F^{s \times 2(n-s)}$ _and_ $\eta_Z = \eta_{A,B} = \psi_n(0,(A,B))$ _as in Lemma_ 8.17. _Then_

(8.19a) $$\left[\psi_r\left(\begin{bmatrix} a & b \\ c & -{}^t a \end{bmatrix}, 0 \right), \zeta_D \right] = 0 \quad \underline{and} \quad \left[\psi_r\left(\begin{bmatrix} a & b \\ c & -{}^t a \end{bmatrix}, 0 \right), \eta_{A,B} \right] = \eta_{A',B'}$$

<u>with</u> $(A',B') = -(A,B)\begin{bmatrix} a & b \\ c & -{}^t a \end{bmatrix}$, _and_

(8.19b) $$[\psi_r(0,\gamma),\zeta_D] = \zeta_{\gamma D + D \cdot {}^t\gamma} \quad \underline{and} \quad [\psi_r(0,\gamma),\eta_Z] = \eta_{\gamma Z}.$$

Comparing Lemma 8.18 with the product formula (8.5b) we conclude:

8.20. <u>Proposition</u>. _Let_ $\psi = \psi_n \oplus \psi_r$ _in the notation of Lemmas_ 8.16 _and_ 8.17. _Then_

$$\psi \colon \mathfrak{p}_{s;2(n-s)}(F) \to \mathfrak{p} \quad \underline{\textit{is a Lie algebra isomorphism,}}$$

and <u>_it restricts to an isomorphism of_</u> $\mathfrak{g}_{s;2(n-s)}(F)$ <u>_onto_</u> $\mathfrak{l}$.

We have proved Theorem 8.6 on the Lie algebra level.

The discussion leading to Lemma 3.11 holds in the present case, and so we have

8.21. *Lemma.* *The isomorphism* $\psi_n : \mathfrak{n}_{s;2(n-s)}(F) \to \mathfrak{p}_n$ *is induced by a Lie group isomorphism* $\phi_n : N_{s;2(n-s)}(F) \to$ *(unipotent radical of* P_E *).*

We need the elementary fact

$$(8.22) \quad \begin{cases} \text{Let} \quad H = \{g: V \to V \quad \text{symplectic:} \quad gE = E \quad \text{and} \quad gE' = E'\}. \quad \text{Then} \\ g \to g|_E \quad \text{maps} \quad H \quad \text{isomorphically onto the general linear group} \\ \text{of} \quad E. \end{cases}$$

This allows us to state

8.23. *Lemma.* *Define* $\phi_r : Sp(n-s;F) \times GL(s;F) \to Sp(n;F)$ *by*

(i) $\phi_r|_{Sp(n-s;F)}$ *is the isomorphism of* $Sp(n-s;F)$ *onto the symplectic group* $\{g \in Sp(n;F): g|_V = 1 \ \underline{and} \ gW = W\}$ *of* W *that is induced by the symplectic isomorphism* $x \to \sum_{\ell=1}^{n-s} e_{s+\ell} x^\ell + \sum_{m=1}^{n-s} e_{n+s+m} x^{n-s+m}$ *of* $F^{2(n-s)}$ *onto* W.

(ii) $\phi_r|_{GL(s;F)}$ *is the isomorphism of* $GL(s;F)$ *onto* $\{g \in Sp(n;F): g|_W = 1, \ gE = E, \ gE' = E'\}$ *specified by* (8.22) *and the isomorphism* $y \to \sum_{\ell=1}^{s} e_\ell y^\ell$ *of* F^s *onto* E.

Then ϕ_r *induces the Lie algebra isomorphism* ψ_r *of Lemma* 8.16, *and* ϕ_r *is an isomorphism of* $Sp(n-s;F) \times GL(s;F)$ *onto* $\{g \in P_E: g\mathfrak{p}_r g^{-1} = \mathfrak{p}_r\}$.

Proof. ϕ_r induces ψ_r by construction, ϕ_r is an isomorphism onto image (ϕ_r), and image $(\phi_r) \subset P_E$. Now image (ϕ_r) is in the P_E-normalizer of image $(\psi_r) = \mathfrak{p}_r$. So we need only verify that image (ϕ_r) contains the P_E-normalizer of $\mathfrak{p}_r$.

The center of $\mathfrak{p}_r$ is the center of $\mathfrak{p}_r^W$, which acts by 0 on V and arbitrary scalars on W. If $g \in P_E$ with $g\mathfrak{p}_r g^{-1} = \mathfrak{p}_r$ now $gV = V$ and $gW = W$. Thus $g = g_V g_W$ where $g_V(v+w) = g(v) + w$ and $g_W(v+w) = v+g_W(w)$ for $v \in V$, $w \in W$. It follows that $g \in \text{image} (\phi_r)$.

q.e.d.

$P = P_E$ has Chevalley semidirect product decomposition $P = P_n \cdot P_r$ where the unipotent radical P_n is the analytic group for its Lie algebra $\mathfrak{p}_n$ and the reductive complement P_r is the normalizer of its Lie algebra $\mathfrak{p}_r$. Now Lemmas 8.16, 8.17, 8.18, 8.21 and 8.23 give us an isomorphism

$$\phi: P_{s;2(n-s)}(F) \to P_E \quad \text{by} \quad .\phi(D,Z;g_W,\gamma) = \phi_n(D;Z) \cdot \phi_r(g_W,\gamma),$$

as required for (8.7a). The restriction fact (8.7c) comes from Lemma 8.23 and the last part of Lemma 8.14. The assertion (8.7b) on the Langlands decomposition is clear at this point, and now the proof of Theorem 8.6 is complete.

q.e.d.

§9. Representations of the Nilradical and the Intermediate Group.

We find the irreducible unitary representations of the groups $N_{s;2t}(F)$, $G_{s;2t}(F)$ and $\tilde{G}_{s;2t}(F)$ of §8. This generally follows the lines of §§4 and 5, except that some representations do not extend to their $G_{s;2t}(F)$-stabilizers; that is why we also consider the metaplectic groups.

In this section we use the notation

(9.1a) $N = N_{s;2t}(F)$, our simply connected nilpotent group;

(9.1b) $n = n_{s;2t}(F) = \text{Sym } F^{s \times s} + F^{s \times 2t}$, its Lie algebra;

(9.1c) $n^* = $ the real linear dual space of n; and

(9.1d) Ad^*: the co-adjoint representation of N on n^*.

A nondegenerate real pairing on n is given by the formula

(9.2a) $\langle (D,Z),(D',Z') \rangle = \text{trace Re } DD' + \text{trace Re } \mathcal{A}(Z,Z')$

and that gives us the rather useful

(9.2b) $n^* = \{ f_{D,Z} \colon (D,Z) \in n \}$ where $f_{D,Z}(D',Z') = \langle (D,Z),(D',Z') \rangle$.

As before we will frequently write f_D for $f_{D,0}$ and g_Z for $f_{0,Z}$. The obvious change of notation transforms the statement and proof of Proposition 4.3 to

9.3. Proposition. *If* $D \in \text{Sym } F^{s \times s}$ *we fix a direct sum decomposition*

(9.4a) $F^{s \times 2t} = \{ Z \in F^{s \times 2t} \colon DZ = 0 \} \oplus F_D^{s \times 2t}$

with the provision: if the matrix D *is semisimple (always the case if*
F = R *or if* s = 1) *then* $F_D^{s \times 2t} = D \cdot F^{s \times 2t}$.

If $f = f_{D,Z} \in n^*$ *then the isotropy algebra* $n_f = \{x \in n: f[x,n] = 0\}$
is

$$(9.4b) \qquad n_f = \text{Sym } F^{s \times s} + \{Z' \in F^{s \times 2t}: DZ' = 0\}$$

and the co-adjoint orbit is the affine subspace

$$(9.4c) \qquad \text{Ad}^*(N) \cdot f = f + \{g_{DZ'}: Z' \in F_D^{s \times 2t}\}.$$

Since $\text{Sym } F^{s \times s}$ always contains nonsingular matrices D, there exist
f with $n_f = \text{Sym } F^{s \times s}$, and so [17]

  9.5. Corollary. $N_{s;2t}(F)$ *has square integrable representations.*
If $f - f_{D,Z} \in n^*$ we denote

$$(9.6a) \qquad \delta_f = \{D' \in \text{Sym } F^{s \times s}: f(D') = 0\}, \quad \text{central ideal in } n;$$

$$(9.6b) \qquad m_f = n/\delta_f \quad \text{quotient algebra and } p: n \to m_f \text{ projection;}$$

$$(9.6c) \qquad \mathfrak{h}_f = p(\text{Sym } F^{s \times s} + F_D^{s \times 2t}) \text{ with } F_D^{s \times 2t} \text{ as in (9.4a); and}$$

$$(9.6d) \qquad \alpha_f = p(\delta_f + \{Z' \in F^{s \times 2t}: DZ' = 0\}).$$

As in §4, $m_f = \mathfrak{h}_f \oplus \alpha_f$, direct sum of ideals. Setting

$$(9.7a) \qquad M_f: \text{ the simply connected nilpotent Lie group for } m_f, \text{ and}$$

$$(9.7b) \qquad H_f \text{ and } A_f: \text{ the analytic subgroups for } \mathfrak{h}_f \text{ and } \alpha_f,$$

we have

(9.7c) $M_f = H_f \times A_f.$

A_f is a vector group, and the possibilities for H_f are

(9.8a) H_f is a Heisenberg group of real dimension ≥ 3 $(D \neq 0, \; t > 0)$,

(9.8b) H_f is a 1-dimensional real vector group $(D \neq 0, \; t = 0)$

(9.8c) H_f is the trivial group $(D = 0)$.

According to the Kirillov Theory, the unitary representation class $[\pi_f] \in \hat{N}$ for $\mathrm{Ad}^*(N) \cdot f$ is given by

(9.9a) $[\pi_f] = [\bar{\pi}_f \cdot p]$ where $[\bar{\pi}_f] \in \hat{M}_f$ is the class for $\bar{f}$ with $f = \bar{f} \cdot p$.

Further, from (9.7c),

(9.9b) $[\bar{\pi}_f] = [\eta_f \otimes \alpha_f]$ where $[\eta_f] \in \hat{H}_f$ and $\alpha_f \in \hat{A}_f.$

Evidently $\alpha_f = e^{ig_Z} : \exp(\delta_f + Z') \to e^{ig_Z(Z')} = e^{i \, \mathrm{trace} \, \mathrm{Re} \, \mathscr{A}(Z,Z')}$ and η_f acts on $p(\exp \mathrm{Sym} \, F^{s \times s})$ by the unitary character

$$e^{if_D} : \exp(\delta_f + D') \to e^{if_D(D')} = e^{i \, \mathrm{trace} \, \mathrm{Re} \, DD'}.$$

According to the three cases of (9.8),

(9.10a) H_f is Heisenberg and $\eta_f \in \hat{H}_f$ has central character $e^{if_D} \neq 1$,

(9.10b) $H_f \cong R^1$ and η_f is the unitary character $e^{if_D} \neq 1$,

(9.10c) H_f and η_f are trivial.

<u>9.11. Theorem.</u> *Let* $N_{s;2t}(F)$. *Then* $\hat{N}$ *consists of the classes* $[\pi_{D,Z}] = [\pi_{f_{D,Z}}]$ *given as follows.*

(9.12a) $D \in \text{Sym } F^{s \times s}$ *and* $Z \in F^{s \times 2t}$ *modulo* $D \cdot F^{s \times 2t}$,

(9.12b) $p: N \to M_f = H_f \times A_f$ *as in* (9.6) *and* (9.7) *with* $f = f_{D,Z}$,

(9.12c) α_f *is the unitary character* e^{ig_Z} *on* A_f,

(9.12d) $[\eta_f] \in \hat{H}_f$ *is the class given by* (9.8) *and* (9.10), *and*

(9.12e) $[\pi_{D,Z}] = [(\eta_f \otimes \alpha_f) \cdot p] \in \hat{N}$.

In addition,

  (i) $[\pi_{D,Z}] = [\pi_{D',Z'}]$ *if and only if* $D = D'$ *and* $Z - Z' \in D \cdot F^{s \times 2t}$,
  (ii)   *the central character of* $[\pi_{D,Z}]$ *restricts to* e^{if_D} *on* $\text{Sym } F^{s \times s}$,
  (iii)   *if* $D \neq 0$ *and* $t > 0$ *then* $[\pi_{D,Z}]$ *is infinite dimensional,*
  (iv)   *if* $D = 0$ *or* $t = 0$ *then* $[\pi_{D,Z}]$ *is a unitary character.*

Finally, the Plancherel measure of $\hat{N}$ *is concentrated on* $\{[\pi_{D,Z}]:$ *the matrix* $D \in \text{Sym } F^{s \times s}$ *is invertible}.*

To study the stabilizers of the $[\pi_{D,Z}]$ in our larger groups, we need

<u>9.13. Lemma.</u> $\text{Sp}(t;F) \times \text{GL}(s;F)$ *acts on* $\mathfrak{n}^*$ *by*
$$(g,\gamma)^{-1} \cdot f_{D,Z} = f_{t_{\gamma \cdot D\gamma}, \, t_{\gamma \cdot Zg}}.$$

This is proved in exactly the same way as Lemma 5.2.

We modify the argument of Lemma 5.4 to prove

<u>9.14. Lemma.</u> *Let* $D \in \text{Sym } F^{s \times s}$ *and* $Z \in F^{s \times 2t}$, *and let* $L_{D,Z}$ *denote the* $\text{Sp}(t;F)$-*stabilizer of* $[\pi_{D,Z}]$. *Then*

(9.15a) $L_{D,Z} = \{g \in \text{Sp}(t;F): Zg - Z \in D \cdot F^{s \times 2t}\}.$

Choose $\gamma \in \mathrm{GL}(s;F)$ *with* ${}^t\gamma \cdot D\gamma$ *semisimple.* *Without changing* $[\pi_{D,Z}]$ *we add an element of* $D \cdot F^{s \times 2t}$ *to* Z *so that* $({}^t\gamma \cdot D\gamma)({}^t\gamma Z) = 0$. *Assuming that normalization of* Z, *let* S_Z *denote the subspace of* F^{2t} *spanned by the columns of* ${}^tZ$. *Then*

$$(9.15\mathrm{b}) \quad L_{D,Z} = \{g \in \mathrm{Sp}(t;F): Zg = Z\} = \{g \in \mathrm{Sp}(t;F): g(x) = x \; \underline{for \; all} \; x \in S_Z\}.$$

and the $\mathrm{Mp}(t;F)$-*stabilizer of* $[\pi_{D,Z}]$ *is*

$$(9.15\mathrm{c}) \quad \tilde{L}_{D,Z} = \varepsilon^{-1} L_{D,Z} = \{\tilde{g} \in \mathrm{Mp}(t;F): (\varepsilon \tilde{g})(x) = x \; \underline{for \; all} \; x \in S_Z\}.$$

Proof. Assertion (9.15a) is just a restatement of (9.4c), in view of the Kirillov Theory.

We have $\gamma \in \mathrm{GL}(s;F)$ diagonalizing the symmetric matrix D, in particular such that ${}^t\gamma \cdot D\gamma$ is semisimple. Assuming that semisimplicity, (9.4a) gives us

$$F^{s \times 2t} = \{{}^t\gamma Z' \in F^{s \times 2t}: ({}^t\gamma D\gamma)({}^t\gamma Z') = 0\} \oplus \{({}^t\gamma D\gamma) \cdot {}^t\gamma F^{s \times 2t}\}.$$

Since γ^{-1} carries co-adjoint orbits to co-adjoint orbits, consequence of the fact that it is an automorphism, (9.4c) lets us normalize Z as required.

Since $\mathrm{Sp}(t;F)$ commutes with $\mathrm{GL}(s;F)$ in $P_{s;2t}(F)$, $L_{D,Z} = L_{{}^t\gamma D\gamma, {}^t\gamma Z}$. In view of our normalization the latter is $\{g \in \mathrm{Sp}(t;F): {}^t\gamma Zg = {}^t\gamma Z\}$. Now $L_{D,Z} = \{g \in \mathrm{Sp}(t;F): Zg = Z\} = \{g \in \mathrm{Sp}(t;F): g(x) = x \text{ on } S_Z\}$.

The assertion on $\tilde{L}_{D,Z}$ follows.

$$q.e.d.$$

Combining Theorem 8.6 with the orthocomplementation argument used in the proof of Proposition 5.6, we obtain the structure of $L_{D,Z}$ and $\tilde{L}_{D,Z}$.

9.16. Proposition. *Let* S *be a subspace of* F^{2t}, c = dim S ∩ $S^{\perp}$, *and* u ⩾ 0 *such that* S ≅ (S ∩ $S^{\perp}$) ⊕ F^{2u}. *Then*

$$\{g \in \mathrm{Sp}(t;F) : g(x) = x \ \textit{for all} \ x \in S\} \cong G_{c;2(t-u-c)}(F)$$

and

$$\{\tilde{g} \in \mathrm{Mp}(t;F) : \varepsilon\tilde{g}(x) = x \ \textit{for all} \ x \in S\} \cong \tilde{G}_{c;2(t-u-c)}(F).$$

Now we have the Mackey little group $N_{s;2t}(F)\cdot L_{D,Z}$ for the class $[\pi_{D,Z}] \in N_{s;2t}(F)^{\wedge}$, and we turn to the extension problem. Here we are forced to consider the Mackey obstruction, whose definition and role are reviewed in §§A.III and A.IV of the Appendix.

If G is a separable locally compact group, then, as in the Appendix, $H^2(G;C')$ denotes cohomology based on Borel cochains $G \times G \rightarrow C'$ with values in the circle group C'. $\pi_1(G)$ denotes the fundamental group of G in the sense of algebraic topology. If A is a locally compact abelian group, then $\hat{A}$ is viewed as its Pontrjagin dual. Our calculation of the Mackey obstruction is based on the following lemma, which basically is contained in Moore ([13], [14]).

9.17. Lemma. *If* G *is a connected semisimple Lie group, then* $H^2(G;C')$ *is naturally isomorphic to* $\pi_1(G)^{\wedge}$. *If* f: J → G *is a continuous homomorphism of connected semisimple Lie groups, then* $f^*: H^2(G;C') \rightarrow H^2(J;C')$ *is the transpose of the induced map* $f_*: \pi_1(J) \rightarrow \pi_1(G)$.

Proof. We follow the terminology of C. C. Moore [13, 14]. Let $\tilde{G} \rightarrow G$ be the universal cover, so $\pi_1(G)$ is its kernel, and consider the restriction-inflation sequence (exact sequence associated to $1 \rightarrow \pi_1(G) \rightarrow \tilde{G} \rightarrow G \rightarrow 1$)

$$\ldots \to H^1(\tilde{G};C') \to H^1(\pi_1(G);C') \to H^2(G;C') \to H^2(\tilde{G};C').$$

$H^2(\tilde{G};C') = \{1\}$ by [14, Proposition 3.4], and $H^1(\tilde{G};C') = (\tilde{G}/[\tilde{G},\tilde{G}])\hat{}$ is

trivial because $\tilde{G}$ is connected and semisimple, so the sequence reduces to

$$1 \to H^1(\pi_1(G);C') \to H^2(G;C') \to 1.$$

As $H^1(\pi_1(G);C')$ is naturally isomorphic to $\pi_1(G)\hat{}$, the assertion follows.

If $f\colon J \to C$ is a continuous homomorphism of connected semisimple Lie

groups, now the associated morphism of restriction-inflation sequences gives

a commutative diagram

$$\pi_1(G)\hat{} \cong H^1(\pi_1(G);C') \cong H^2(G;C')$$

$$\downarrow {}^t(f_*) \qquad\qquad \downarrow (f_*)^* \qquad\qquad \downarrow f^*$$

$$\pi_1(J)\hat{} \cong H^1(\pi_1(J);C') \cong H^2(J;C')$$

which identifies f^* to ${}^t(f_*)$.

$$q.e.d.$$

We apply Lemma 9.17 to symplectic groups. The first assertion is

standard.

9.18. Lemma.

1) $H^2(Sp(t,C);C') = \{1\}$, _and_ $H^2(Sp(t;R);C')$ _is a circle group for_

   $t \geqslant 1$.

2) _If_ $r, t \geqslant 1$ _and_ $f\colon Sp(t;R) \to Sp(rt;R)$ _by_ $f\begin{bmatrix} a & b \\ c & d \end{bmatrix} = \begin{bmatrix} a\otimes I & b\otimes I \\ c\otimes I & d\otimes I \end{bmatrix}$,

then $f^*\colon H^2(Sp(rt;R);C') \to H^2(Sp(t;R);C')$ _is an isomorphism_.

3) _If_ $1 \leqslant u \leqslant t$ _and_ $W \cong F^{2u}$ _is a nonsingular subspace of_ F^{2t} ,

and if $\varphi\colon Sp(u;R) \to Sp(t;R)$ _is the corresponding inclusion of the_ *symplectic group of* W, *then* $\varphi^*\colon H^2(Sp(t;R);C') \to H^2(Sp(u;R);C')$ *is*

an isomorphism.

Proof. $Sp(t;C)$ is simply connected, so $H^2(Sp(t;C);C') = \{1\}$ by

Lemma 9.17. $Sp(t;R)$ has maximal compact subgroup $U(t)$. If $t \geqslant 1$ now

$\pi_1(Sp(t;R)) = \pi_1(U(t))$, which is infinite cyclic, so $H^2(Sp(t;R):C') \cong$ (in-

finite cyclic)$^{\wedge} \cong C'$ by Lemma 9.17.

In (2), the restriction $f\colon U(t) \to U(rt)$ to maximal compact subgroups,

gives an isomorphism of the center of $U(t)$ onto the center of $U(rt)$. It

follows that

$$f_*\colon \{\pi_1(Sp(rt;R)) = \pi_1(U(rt))\} \to \{\pi_1(U(t)) = \pi_1(Sp(t;R))\}$$

is an isomorphism. Now $f^*\colon H^2(Sp(rt;R);C') \to H^2(Sp(t;R);C')$ is an isomor-

phism by Lemma 9.17.

In (3), the restriction $\varphi\colon U(u) \to U(t)$ to maximal compact subgroups is

$\varphi(k) = \begin{pmatrix} k & 0 \\ 0 & I \end{pmatrix}$. If $u = t$ there is nothing to prove, so let us suppose

$u < t$. If $u = t - 1$ we consider the principal $U(u)$-bundle,

$$U(u+1) \to U(u+1)/U(u) = S^{2u+1},\ \text{unit sphere in } C^{u+1}.$$

Part of its homotopy sequence is

$$\{1\} = \pi_2(S^{2u+1}) \to \pi_1(U(u)) \to \pi_1(U(u+1)) \to \pi_1(S^{2u+1}) = \{1\}\ ,$$

so $\varphi_*\colon \pi_1(U(u)) \to \pi_1(U(u+1))$ is an isomorphism. If $u < t-1$, we factor φ

into $U(u) \overset{\alpha}{\to} U(t-1) \overset{\beta}{\to} U(t)$. By induction on t for fixed u,

$\alpha_*\colon \pi_1(U(u)) \to \pi_1(U(t-1))$ is an isomorphism. By the case $u = t-1$,

$\beta_*\colon \pi_1(U(t-1)) \to \pi_1(U(t))$ is an isomorphism. Now $\varphi_*\colon \pi_1(U(u)) \to \pi_1(U(t))$

is an isomorphism, and so $\varphi^*\colon H^2(Sp(t;R);C') \to H^2(Sp(u;R);C')$ is an iso-

morphism as in (2).

$$q.e.d.$$

Now we can settle the question of extending $[\pi_{D,Z}]$ from $N_{s;2t}(F)$ to $N_{s;2t}(F) \cdot L_{D,Z}$.

9.19. Theorem. *Let* $[\pi_{D,Z}] \in N_{s;2t}(F)^\wedge$ *as in Theorem* 9.11 *with the normalization of Lemma* 9.14. *Let* S_Z *denote the subspace of* F^{2t} *spanned by the columns of* ^tZ, $c = \dim(S_Z \cap S_Z^\perp)$, *and* $u \geqslant 0$ *such that* $S_Z \cong (S_Z \cap S_Z^\perp) \oplus F^{2u}$, *so that*

$$L_{D,Z} \cong G_{c;2(t-u-c)}(F) = N_{c;2(t-u-c)}(F) \cdot Sp(t-u-c;F).$$

Then $\pi_{D,Z}$ *extends to a unitary representation* $\tilde{\pi}_{D,Z}$ *of* $N_{s;2t}(F) \cdot L_{D,Z}$ *on the same Hilbert space if, and only if, one or more of the following conditions* (9.20) *holds.*

(9.20a) $F = C$

(9.20b) $\pi_{D,Z}$ *is a unitary character, that is* $D = 0$ *or* $t = 0$.

(2.20c) $L_{D,Z}$ *is nilpotent, that is* $t = u + c$.

If the conditions (9.20) *all fail, then the Mackey obstruction to extension is an element* $k(\pi_{D,Z}) \in H^2(L_{D;Z};C')$ *of order* 2, *obtained as follows:*

(9.12a) $h: L_{D,Z} \to Sp(t-u-c;R)$ *is projection with kernel* $N_{c;2(t-u-c)}(R)$,

(9.21b) ν *is the element of order   in the circle group*
          $H^2(Sp(t-u-c;R);C')$,

(9.21c) $k(\pi_{D,Z}) = h^*(\nu)$.

{*In that case,* $\pi_{D,Z}$ *only extends to a cocycle representation* $\tilde{\pi}_{D,Z}$ *of* $N_{s;2t}(R) \cdot L_{D,Z}$ *on the same Hilbert space*} .

9.22. Corollary. *If* $[\pi_{D,Z}] \in N_{s;2t}(F)^\wedge$, *then it extends to a unitary representation class* $[\tilde{\pi}_{D,Z}]$ *of its* $\tilde{G}_{s;2t}(F)$-*normalizer* $N_{s;2t}(F) \cdot \tilde{L}_{D,Z}$ *on*

the same Hilbert space.

Proof of Corollary from Theorem. The Mackey obstruction $k(\pi_{D,Z})$ to extending $\pi_{D,Z}$ to $N_{s;2t}(F) \cdot L_{D,Z}$ is a cohomology class of order 1 or 2, annihilated by the pullback from $\tilde{L}_{D,Z} \to L_{D,Z}$. Thus the obstruction to extending $\pi_{D,Z}$ to $N_{s;2t}(F) \cdot \tilde{L}_{D,Z}$ is trivial.

Remark. Following the proof of Theorem 9.19, the reader will see the Corollary 9.22 is a rather general fact, for the key point is the metaplectic property discussed just after (8.5).

Proof of Theorem. Let $f = f_{D,Z}$. By (9.15b), the unitary character $\alpha_f \in \hat{A}_f$ is $L_{D,Z}$-stable, so the formula $\tilde{\alpha}_f(a,g) = \alpha_f(a)$ extends it to a unitary character $\tilde{\alpha}_f$ on $A_f \cdot L_{D,Z}$.

If H_f is commutative, then $\eta_f \in \hat{H}_f$ similarly extends to a unitary character $\tilde{\eta}_f(h,g) = \eta_f(h)$ on $H_f \cdot L_{D,Z}$.

Now suppose H_f noncommutative. Then H_f is a Heisenberg group of real dimension $2v + 1$ where $v = (\dim_R F)(\text{rank } D)t > 0$. We view H_f as the quotient M_f/A_f so that $L_{D,Z}$ acts on it by automorphisms. Denote Mackey obstructions

$$k(\eta_f) \in H^2(L_{D,Z};C'): \text{ to extending } \eta_f \text{ to } H_f \cdot L_{D,Z}$$

and

$$k'(\eta_f) \in H^2(Sp(t;F);C'): \text{ to extending } \eta_f \text{ to } H_f \cdot Sp(t;F).$$

Then $k(\eta_f)$ is the restriction $j^* \cdot k'(\eta_f)$ where $j: L_{D,Z} \to Sp(t;F)$ is inclusion.

If $F = C$, then $k'(\eta_f) = 1$ by Lemma 9.18 (1), so $k(\eta_f) = 1$. Then η_f extends to a unitary representation $\tilde{\eta}_f$ of $H_f \cdot L_{D,Z}$ on the same Hilbert space.

Now H_f is noncommutative and $F = R$. Let $r = \text{rank } D$, so H_f is isomorphic to the Heisenberg group $N_{1;2rt}(R)$ of dimension $2rt + 1$. View $Sp(rt;R)$ as the group of all automorphisms of H_f that act trivially on the center. The inclusion $Sp(t;R) \to Sp(rt;R)$ is the inclusion of Lemma 9.18 (2). One knows [21] that the Mackey obstruction to extending on infinite dimensional class $[\eta] \in N_{1;2v}(R)$ to $N_{1;2v}(R) \cdot Sp(v;R)$ is an element of order 2, that is η only extends when $Sp(v;R)$ is replaced by a covering group of even multiplicity. Let $k''(\eta_f)$ denote the Mackey obstruction to extending η_f to $H_f \cdot Sp(rt;R)$. It pulls back to $k'(\eta_f)$ under the inclusion $Sp(t;R) \to Sp(rt;R)$. In view of Lemma 9.18 (2), now

(9.23) $k'(\eta_f)$ is the element of order 2 in the circle $H^2(Sp(t;R);C')$.

Write V for the subgroup $N_{c;2(t-c-u)}(R)$ of $L_{D,Z}$. Then $H_f \cdot V$ is a connected simply connected nilpotent Lie group. The Kirillov theory tells us that η_f extends to a unitary representation η'_f of $H_f \cdot V$ on the representation space of η_f. If we denote Mackey obstruction

(9.24a) $k^V(\eta_f) \in H^2(Sp(t-u-c;R);C')$: to extending η'_f to $H_f \cdot L_{D,Z}$,

then it follows that

(9.24b) $k(\eta_f) = h^* \cdot k^V(\eta_f)$ where $h: L_{D,Z} \to L_{D,Z}/V = Sp(t-u-c;R)$.

Let $\varphi: Sp(t-u-c;R) \to Sp(t;R)$, inclusion. If $u + c < t$, Lemma 9.18 (3) says that $\varphi^*: H^2(Sp(t;R);C') \to H^2(Sp(t-u-c;R);C')$ is an isomorphism. But evidently $k^V(\eta_f) = \varphi^* \cdot k'(\eta_f)$. In view of (9.23) and (9.24), now

(9.25) $\begin{cases} \textit{if } u + c < t, \textit{ so that } H^2(Sp(t-u-c;R);C') \textit{ is a circle} \\ \textit{group, then } k(\eta_f) = h^*(\nu) \textit{ where } \nu \textit{ is its element of} \\ \textit{order } 2. \end{cases}$

We summarize to this point. We have shown that (9.20) is a necessary and sufficient condition that η_f extend to a unitary representation $\tilde{\eta}_f$ of $H_f \cdot L_{D,Z}$ on the same space, and when (9.20) fails we have shown that $k(\eta_f) \in H^2(L_{D,Z};C')$ is the element of order 2 given by (9.25).

Let $\mu \in H^2(L_{D,Z};C')$. Suppose that η_f extends to a μ-representation $\tilde{\eta}_f$ of $H_f \cdot L_{D,Z}$ on the same space. Let $\tilde{p}: N_{s;2t}(F) \cdot L_{D,Z} \to (H_f \times A_f) \cdot L_{D,Z}$ be the obvious extension of $p: N_{s;2t}(F) \to H_f \times A_f$. Then $\tilde{\pi}_{D,Z} = (\tilde{\eta}_f \otimes \tilde{\alpha}_f) \cdot p$ extends $\pi_{D,Z}$ to a μ-representation of $N_{s;2t}(F) \cdot L_{D,Z}$ on the same space. In particular, $k(\pi_{D,Z}) = \mu = k(\eta_f)$. This completes the proof of Theorem 9.19.   *q.e.d.*

In general we extend the covering $\varepsilon: Mp(b;F) \to Sp(b;F)$ to

(9.26a) $\qquad \varepsilon: \tilde{G}_{a;2b}(F) \to G_{a;2b}(F)$ by $\varepsilon(D,Z;\tilde{g}) = (D,Z;\varepsilon\tilde{g})$,

and we view

(9.26b) $\qquad G_{a;2b}(F)^\wedge = \{[\pi] \in \tilde{G}_{a;2b}(F)^\wedge: \pi \text{ has form } \bar{\pi} \cdot \varepsilon\}$.

If $[\pi_{D,Z}] \in N_{s;2t}(F)^\wedge$ satisfies (9.20), then the irreducible unitary representation classes that restrict to a multiple of $[\pi_{D,Z}]$ are

(9.27a) $\quad$ for $N_{s;2t}(F) \cdot \tilde{L}_{D,Z}$: all $[\tilde{\pi}_{D,Z} \otimes \gamma]$ with $[\gamma] \in (\tilde{L}_{D,Z})^\wedge$

and

(9.27b) $\quad$ for $N_{s;2t}(F) \cdot L_{D,Z}$: all $[\tilde{\pi}_{D,Z} \otimes \gamma]$ with $[\gamma] \in (L_{D,Z})^\wedge$.

On the other hand, if (9.20) fails for $[\pi_{D,Z}]$, then the irreducible unitary representation classes that restrict to a multiple of $[\pi_{D,Z}]$ are

(9.28a)   for $N_{s;2t}(R)\cdot\tilde{L}_{D,Z}$: all $[\tilde{\pi}_{D,Z} \otimes \gamma]$ with $[\gamma] \in (\tilde{L}_{D,Z})^\wedge$

but

(9.28b)   for $N_{s;2t}(R)\cdot L_{D,Z}$: all $[\tilde{\pi}_{D,Z} \otimes \gamma]$ with $[\gamma] \in (\tilde{L}_{D,Z})^\wedge - (L_{D,Z})^\wedge$.

Now we apply the Mackey little group method. Given $[\pi_{D,Z}] \in N_{s;2t}(F)^\wedge$, the associated representation classes for $\tilde{G}_{s;2t}(F) = N_{s;2t}(F)\cdot Mp(t;F)$ are the

(9.29a) $[\pi_{D,Z,\gamma}] = [\mathrm{Ind}_{N_{s;2t}(F)\cdot\tilde{L}_{D,Z}\uparrow\tilde{G}_{s;2t}(F)}(\tilde{\pi}_{D,Z} \otimes \gamma)]$

where

(9.29b) $[\gamma] \in (\tilde{L}_{D,Z})^\wedge$ extended as usual by $\gamma(D',Z';\tilde{g}) = \gamma(\tilde{g})$.

The associated representation classes for $G_{s;2t}(F) = N_{s;2t}(F)\cdot Sp(t;F)$ are

(9.30a)   if (9.20) holds: those $[\pi_{D,Z,\gamma}]$ with $[\gamma] \in (L_{D,Z})^\wedge$,

(9.30b)   if (9.20) fails: those $[\pi_{D,Z,\gamma}]$ with $[\gamma] \notin (L_{D,Z})^\wedge$.

In view of Theorem 9.11, every irreducible unitary representation of $\tilde{G}_{s;2t}(F)$ is equivalent to a $\pi_{D,Z,\gamma}$ specified in (9.29), and every irreducible unitary representation of $G_{s;2t}(F)$ is equivalent to a $\pi_{D,Z,\gamma}$ specified in (9.30). Assume the normalization of Lemma 9.14 with D and D' semisimplified by the same element of $GL(s;F)$. Then $[\pi_{D,Z,\gamma}] = [\pi_{D',Z',\gamma'}]$ if and only if there is an element $g \in Sp(t;F)$ such that

(9.31) $D = D'$, $Z' = Zg^{-1}$ and $\tilde{g}' \to \gamma(g\tilde{g}'g^{-1})$ is equivalent to γ'.

To be explicit about (9.31), we note that $Z' \in Z \cdot Sp(t;F)$ just when an element of $Sp(t;F)$ carries each column of ^tZ to the corresponding column of $^tZ'$. Write $Z = (X,Y)$ and $Z' = (X',Y')$ in $t \times t$ blocks, and consider the bilinear form

$$(9.32a) \qquad \mathcal{B}: F^{s \times 2t} \times F^{s \times 2t} \to F^{s \times s} \quad \text{by} \quad B(Z,Z') = X \cdot {}^tY' - Y \cdot {}^tX'$$

such that $\mathcal{A}(Z,Z') = \frac{1}{2}\{\mathcal{B}(Z,Z') - \mathcal{B}(Z',Z)\}$. Then

$$(9.32b) \qquad Z' \in Z \cdot Sp(t;F) \Leftrightarrow \text{rank } Z = \text{rank } Z' \quad \text{and} \quad \mathcal{B}(Z,Z) = \mathcal{B}(Z',Z').$$

The integers $c = \dim(S_Z \cap S_Z^\perp)$ and $u = \frac{1}{2}\{\dim S_Z - \dim(S_Z \cap S_Z^\perp)\}$ associated to the column span S_Z of ^tZ in Proposition 9.16, are related by

$$(9.32c) \qquad \text{rank } Z = c + 2u \quad \text{and} \quad \text{rank } \mathcal{B}(Z,Z) = 2u.$$

Recalling Lemma 9.13 and Proposition 9.16, we formulate the preceeding results and discussion as follows.

$\underline{9.33. \text{ Theorem.}}$ $\tilde{G}_{s;2t}(F)^{\wedge}$ *is the disjoint union of non-empty subsets* $\tilde{S}(c,u;D)$. $D \in \text{Sym } F^{s \times s}$ *and* c, u *non-negative integers such that*

$$(*) \qquad\qquad c + 2u + \text{rank } D \leqslant s \quad \underline{and} \quad c + u \leqslant t,$$

given as follows. Choose $\beta \in GL(s;F)$ *with* $^t\beta \cdot D\beta$ *semisimple. Let* $\mathcal{A}_{u,D}$ *denote the space of all* $s \times s$ *antisymmetric matrices* A *over* F *such that* $(^t\beta D \beta)(^t\beta A \beta) = 0$ *and* rank $A = 2u$. *Then* $\tilde{S}(c,u,D)$ *is parametrized by*

$$\mathcal{A}_{u,D} \times \tilde{G}_{c;2(t-u-c)}(F)^{\wedge}$$

under

$$(A, [\gamma]) \leftrightarrow [\pi_{D,Z,\gamma}]$$

where

$$Z \in F^{s \times 2t} \ \underline{with} \ (^t\beta \cdot D\beta)(^t\beta \cdot Z) = 0, \ \mathcal{B}(Z,Z) = A \ \underline{and} \ \text{rank } Z = c + 2u,$$

and where

$$\tilde{L}_{D,Z} \ \text{is identified with its isomorph} \ \tilde{G}_{c;2(t-u-c)}(F).$$

{*In other words,* $\tilde{S}(c,u;D)$ *is the set of all* $[\pi_{D,Z,\gamma}]$ *that satisfy* (9.32c).}

Proof. Let $\tilde{S}(c,u;D)$ denote the set of all $[\pi_{D,Z,\gamma}]$ in $G_{s;2t}(F)^\wedge$ with $Z \in F^{s \times 2t}$, Z normalized as in Lemma 9.14 using $\beta \in GL(s;F)$, rank $Z = c + 2u$ and rank $\mathcal{B}(Z,Z) = 2u$. We must check that $\tilde{S}(c,u;D)$ is non-empty just when (*) holds, that $\tilde{G}_{s;2t}(F)^\wedge$ is disjoint union of the $\tilde{S}(c,u;D)$, and that the parameterization is correct.

Let $[\pi] \in \tilde{G}_{s;2t}(F)^\wedge$. Then $[\pi] = [\pi_{D,Z,\gamma}]$ where $D \in \text{Sym } F^{s \times s}$, $Z \in F^{s \times 2t}$ with Z normalized as in Lemma 9.14 using $\beta \in GL(s;F)$, and $[\gamma] \in (\tilde{L}_{D,Z})^\wedge$. There is no choice in D, but Z is free over $Z \cdot Sp(t;F)$, which just means that $u = \frac{1}{2} \text{rank } \mathcal{B}(Z,Z)$ and $c = \text{rank } Z - 2u$ are fixed. Thus $[\pi_{D,Z,\gamma}] \to (\mathcal{B}(Z,Z), [\gamma])$ is a one to one map of $\tilde{S}(c,u;D)$ into $\mathcal{A}_{u,D} \times \tilde{G}_{c;2(t-u-c)}(F)^\wedge$, and $\tilde{G}_{s;2t}(F)^\wedge$ is disjoint union of the non-empty $\tilde{S}(c,u;D)$.

Given $[\pi_{D,Z,\gamma}]$ as above, the subspace $S_Z \subset F^{2t}$ has $c = \dim(S_Z \cap S_Z^\perp)$ and $S_Z \cong (S_Z \cap S_Z^\perp) \oplus F^{2u}$. Thus $c + u \leq t$. Also, $DZ = 0$ gives rank $Z \leq s - \text{rank } D$, that is $c + 2u + \text{rank } D = \text{rank } Z + \text{rank } D \leq s$. This proves (*) necessary for $\tilde{S}(c,u;D)$ to be non-empty.

Conversely fix $D \in \text{Sym } F^{s \times s}$ and let $c, u \geqslant 0$ satisfying (*). We have $\gamma \in GL(s;F)$ with $\gamma D \cdot {}^t\gamma = \begin{bmatrix} D' & 0 \\ 0 & 0 \end{bmatrix}$ where $D' \in \text{Sym } F^{r \times r}$ nonsingular and diagonal. If $A \in \mathscr{A}_{u,D}$ now $(\gamma D \cdot {}^t\gamma)({}^t\gamma^{-1} \cdot A \cdot \gamma^{-1}) = 0$, so ${}^t\gamma^{-1} \cdot A \cdot \gamma^{-1} = \begin{bmatrix} 0 & 0 \\ 0 & A' \end{bmatrix}$ with $A' \in F^{(s-r) \times (s-r)}$ antisymmetric of rank $2u$. Because of (*), F^{2t} has a subspace S of dimension $c + 2u$ with $c = \dim(S \cap S^\perp)$ and S is (over)-spanned by vectors $\{z_1, \ldots, z_{s-r}\}$ such that $A' = (\alpha(z_\ell, z_m))$. Let $Z' \in F^{(s-r) \times 2t}$, its ℓ-th row being the transpose of z_ℓ, and define $Z = {}^t\gamma \cdot \begin{pmatrix} 0 \\ Z' \end{pmatrix}$. Then

$$ DZ = \gamma^{-1} \cdot \begin{bmatrix} D' & 0 \\ 0 & 0 \end{bmatrix} \cdot {}^t\gamma^{-1} \cdot {}^t\gamma \begin{pmatrix} 0 \\ Z' \end{pmatrix} = 0 $$

and

$$ \mathscr{B}(Z,Z) = \mathscr{B}\left({}^t\gamma \begin{pmatrix} 0 \\ Z' \end{pmatrix}, \; {}^t\gamma \begin{pmatrix} 0 \\ Z' \end{pmatrix} \right) = {}^t\gamma \begin{bmatrix} 0 & 0 \\ 0 & A' \end{bmatrix} \gamma = A. $$

As rank $Z = \dim S = c + 2u$, now $[\pi_{D,Z,\gamma}]$ exists and is contained in $\tilde{S}(c,u;D)$, corresponding to parameter value $(A, [\gamma])$ there. We have shown that (*) is sufficient for $\tilde{S}(c,u;D)$ to be non-empty, and that $[\pi_{D,Z,\gamma}] \to (A, [\gamma])$ maps $\tilde{S}(c,u;D)$ onto $\mathscr{A}_{u,D} \times \tilde{G}_{c;2(t-c-u)}(F)^\wedge$ when (*) holds. This completes the proof of Theorem 9.33.   *q.e.d.*

9.34. **Corollary**. $G_{s;2t}(F)^\wedge$ *is the disjoint union of non-empty subsets* $S(c,u;D)$, $D \in \text{Sym } F^{s \times s}$ *and* c, u *integers* $\geqslant 0$ *such that*

(*) $c + 2u + \text{rank } D \leqslant s$ *and* $c + u \leqslant t$,

given as follows. Let $\mathscr{A}_{u,D}$ *be as in Theorem* 9.33. *Then* $S(c,u,D)$ *is* *parameterized by*

$$\mathcal{A}_{u,D} \times G_{c;2(t-u-c)}(F)^{\wedge} \quad \textit{if} \quad (9.20) \quad \textit{holds}$$

$$\mathcal{A}_{u,D} \times \{\widetilde{G}_{c;2(t-u-c)}(F)^{\wedge} - G_{c;2(t-u-c)}(F)^{\wedge}\} \quad \textit{if} \quad (9.20) \quad \textit{fails}$$

under

$$(A,[\gamma]) \leftrightarrow [\pi_{D,Z,\gamma}] \quad \textit{as in Theorem} \quad 9.33.$$

Proof. Combine (9.30) with Theorem 9.33. *q.e.d.*

9.35. Comment. In the manner sketched at the end of §5, Theorem 9.33 and Corollary 9.34 give a complete description of the unitary duals $G_{s;2t}(F)^{\wedge}$ and $\widetilde{G}_{s;2t}(F)^{\wedge}$ subject to knowing the $Mp(\ell;F)^{\wedge}$ for $0 < \ell \leqslant t$.

§10. Representations of the Maximal Parabolic Subgroups

We now combine the results of §§1 and 9, using the Mackey little-group method to write out the irreducible unitary representations of the maximal parabolic subgroups of the symplectic and metaplectic groups. Recall that those parabolic subgroups are the

$$P_E = \{g \in Sp(n;F) \colon gE = E\} \text{ and } \tilde{P}_E = \{\tilde{g} \in Mp(n;F) \colon (\varepsilon\tilde{g})E = E\}$$

where E is a totally isotropic subspace of dimension $s > 0$ in F^{2n}. They are isomorphic to

$$P_{s;2(n-s)}(F) = G_{s;2(n-s)}(F)\cdot GL(s;F) \quad \text{and} \quad \tilde{P}_{s;2(n-s)}(F) = \tilde{G}_{s;2(n-s)}(F)\cdot GL(s;F)$$

according to Theorem 8.6. The procedure for determining their unitary duals is valid for their maximal unimodular subgroups

$$P'_{s;2(n-s)}(F) = G_{s;2(n-s)}(F)\cdot GL'(s;F) \quad \text{and} \quad \tilde{P}'_{s;2(n-s)}(F) = \tilde{G}_{s;2(n-s)}(F)\cdot GL'(s;F)$$

and also for other classes in which $GL(s;F)$ is cut down to a reasonable sub-group.

For convenience and reference to §9 we set $t = n-s$.

10.1. Lemma. *The* $GL(s;F)$*-stabilizer of a class* $[\pi_{D,Z,\gamma}] \in \tilde{G}_{s;2t}(F)\hat{}$ *defined in* (9.29) *is*

$$M_{D,Z} = \{\beta \in GL(s;F) \colon {}^t\beta\cdot D\beta = D \text{ \underline{and}}$$
$$\phantom{M_{D,Z} = \{} {}^t\beta\cdot Z - Z\mathbf{g} \in D\cdot F^{s\times 2t} \text{ \underline{for some}} \ \mathbf{g} \in Sp(t;F)\}.$$

Let $d = \operatorname{rank} D$. *If* $F = R$, *let* $(d',d'',s-d)$ *denote the "signature" of*

D. _Recall_ $\mathcal{B}$ _from_ (9.32) _and split_ s-d = e' + 2e'' _where_ $\mathcal{B}(Z,Z)$ _has rank_ 2e''. _Then_

(10.2a) $M_{D,Z} \cong F^{d \times (e' + 2e'')} \cdot \{H \times K\}$ _where_

(10.2b)   H = O(d;C) _if_ F = C, H = O(d',d'') _if_ F = R, _and_

(10.2c) K = {(x;a,b) $\in L_{e',2e''}(F)$: b $\in$ Sp(e'';F)} $\subset$ GL(e' + 2e'';F).

{_In particular_ $K \cong P_{e';2e''}(F)/\mathrm{Sym}\, F^{e' \times e'} = P_{e';2e''}/(\underline{center\ of}\ N_{e';2e''}(F))$}.

Proof: GL(s;F) centralizes Mp(t;F) in $\tilde{P}_{s;2t}(F)$, so we combine Lemma 9.13 with (9.4c) and the equivalence criterion (9.31) to see $M_{D,Z} = \{\beta \in \mathrm{GL}(s;F): {}^t\beta \cdot D\beta = D$ and ${}^t\beta \cdot Z - Zg \in D \cdot F^{s \times 2t}$ where $g \in \mathrm{Sp}(t;F)\}$.

We may replace $[\pi_{D,Z,\gamma}]$ by a class in its GL(s;F)-orbit and assume that $D = \begin{pmatrix} D' & 0 \\ 0 & 0 \end{pmatrix} \in F^{s \times s}$ where $D' \in F^{d \times d}$ is $\begin{pmatrix} I_{d'} & 0 \\ 0 & -I_{d''} \end{pmatrix}$ if F = R, I if F = C. Express $\beta \in$ GL(s;F) in block form $\begin{pmatrix} a & b \\ c & d \end{pmatrix}$. Then

$$ {}^t\beta \cdot D\beta = D \Leftrightarrow {}^t a \cdot D' \cdot a = D' \quad \text{and} \quad b = 0. $$

DZ = 0 says $Z = \begin{pmatrix} 0 \\ Z' \end{pmatrix}$ with $Z' \in F^{(s-d) \times 2t}$, and $D \cdot F^{s \times 2t}$ consists of all matrices $\begin{pmatrix} Z'' \\ 0 \end{pmatrix}$ with $Z'' \in F^{d \times 2t}$. Assume ${}^t\beta \cdot D\beta = D$ and calculate

$$ {}^t\beta \cdot Z - Zg = \begin{bmatrix} {}^t c \cdot Z' \\ {}^t d \cdot Z' - Z'g \end{bmatrix}; \quad \text{so} \quad {}^t\beta \cdot Z - Zg \in D \cdot F^{s \times 2t} \Leftrightarrow {}^t d \cdot Z' = Z'g. $$

That gives us

$$ M_{D,Z} \cong \left\{ \begin{bmatrix} a & 0 \\ c & d \end{bmatrix} : {}^t a \cdot D' a = D' \quad \text{and} \quad {}^t d \cdot Z' \in Z' \cdot \mathrm{Sp}(t;F) \right\}. $$

Under $\beta \to {}^t\beta^{-1}$ now

$$M_{D,Z} \cong \left\{ \begin{bmatrix} a & b \\ 0 & d \end{bmatrix} : aD' \cdot {}^{t}a = D' \quad \text{and} \quad dZ' \in Z' \cdot Sp(t;F) \right\}.$$

Under the isomorphism of Lemma 1.4 now $M_{D,Z} \cong F^{d \times (s-d)} \cdot \{H \times K\}$ where

$H = \{a \in GL(d;F) : aD' \cdot {}^{t}a = D'\}$ and $K = \{d \in GL(s-d;F) : dZ' \in Z' \cdot Sp(t;F)\}$.

A glance back at the form of D' shows that

$$H = O(d;C), \quad \text{complex orthogonal group, if} \quad F = C,$$

and

$$H = O(d',d''), \quad \text{indefinite real orthogonal group, if} \quad F = R.$$

A glance back to (9.32) tells us that

$$K = \{d \in GL(s-d;F) : d \cdot \mathcal{B}(Z',Z') \cdot {}^{t}d = \mathcal{B}(Z',Z')\}.$$

Write $F^{s-d} = F^{e'} \oplus F^{2e''}$ where $F^{e'}$ is the null space of the antisymmetric bilinear form with matrix $\mathcal{B}(Z',Z')$. Following the isomorphism of Lemma 1.4 we see

$$K \cong \left\{ \begin{pmatrix} \sigma & \chi\tau \\ 0 & \tau \end{pmatrix} : \sigma \in GL(e';F) \quad \text{and} \quad \tau \in Sp(e'';F) \right\},$$

that is

$$K \cong \{(\chi;\sigma,\tau) \in L_{e',2e''}(F) : \tau \in Sp(e'';F)\}.$$

This completes the proof of Lemma 10.1.

q.e.d.

    10.3. Lemma. *If* $[\pi_{D,Z,\gamma}] \in \tilde{G}_{s;2t}(F)^{\wedge}$, *then* $\pi_{D,Z,\gamma}$ *extends to a unitary representation* $\tilde{\pi}_{D,Z,\gamma}$ *of* $\tilde{G}_{s;2t}(F) \cdot M_{D,Z}$ *on the same space.*

    Proof. $f_{D,Z}$ has $Mp(t;F) \times GL(s;F)$ stabilizer $Q_{D,Z} = \{(\tilde{g},\beta) : {}^{t}\beta \cdot D\beta = D \text{ and } {}^{t}\beta \cdot Z \cdot \varepsilon(\tilde{g}) = Z\}$, and $[\pi_{D,Z}]$ extends to a class $[\pi'_{D,Z}] \in \{N_{s;2t}(F) \cdot Q_{D,Z}\}^{\wedge}$ by the following modification of Theorem 9.19. Instead of considering the obstruction to extending η_f to the subgroup $H_f \cdot Sp(t;F)$ of $H_f \cdot Sp(rt;F)$, we consider the obstruction to extending η_f to $H_f \cdot Q_{D,Z} \subset H_f \cdot Mp(rt;F)$, and the latter vanishes because η_f extends to $H_f \cdot Mp(rt;F)$.

Let $R_{D,Z}$ denote the projection of $Q_{D,Z}$ to $Mp(t;F)$. Then $R_{D,Z} = \{\tilde{g} \in Mp(t;F) : \epsilon(\tilde{g})S_Z = S_Z\}$ in the notation (9.15), with $\tilde{L}_{D,Z} = \{\tilde{g} \in Mp(t;F) : \epsilon(\tilde{g})$ is the identity of $S_Z\}$. Let V_γ denote the $R_{D,Z}$-stabilizer of $[\gamma] \in (\tilde{L}_{D,Z})^\wedge$. By induction on s, $[\gamma]$ extends to a class $[\gamma'] \in \hat{V}_\gamma$.

Now $[\tilde{\pi}_{D,Z} \otimes \gamma]$ extends from $N_{s;2t}(F) \cdot \tilde{L}_{D,Z}$ to a unitary equivalence class $[\omega]$ on $W = N_{s;2t}(F) \cdot \{Q_{D,Z} \cap (V_\gamma \times GL(s;F))\}$ by $\omega(D',Z';\tilde{g},\beta) = \pi'_{D,Z}(D',Z',\tilde{g},\beta) \otimes \gamma'(\tilde{g})$. We set

$$\tilde{\pi}_{D,Z,\gamma} = \text{Ind}_{W \uparrow \tilde{G}_{s;2t}(F) \cdot M_{D,Z}}(\omega).$$

As $Q_{D,Z} \cap \{Mp(t,F) \times 1\} = \tilde{L}_{D,Z}$, we have $W \cap \tilde{G}_{s;2t}(F) = N_{s;2t}(F) \cdot \tilde{L}_{D,Z}$. So the base space of the fibration used in inducing ω up to $\tilde{\pi}_{D,Z,\gamma}$ is

$$\tilde{G}_{s;2t}(F) \cdot M_{D,Z}/W \approx \tilde{G}_{s;2t}(F)/N_{s;2t}(F) \cdot \tilde{L}_{D,Z}.$$

Thus $\tilde{\pi}_{D,Z,\gamma}$ has $\tilde{G}_{s;2t}(F)$-restriction

$$\text{Ind}_{N_{s;2t}(F) \cdot \tilde{L}_{D,Z} \uparrow \tilde{G}_{s;2t}(F)} \left(\omega|_{N_{s;2t}(F) \cdot \tilde{L}_{D,Z}} \right)$$

$$= \text{Ind}_{N_{s;2t}(F) \cdot \tilde{L}_{D,Z} \uparrow \tilde{G}_{s;2t}(F)} (\tilde{\pi}_{D,Z} \otimes \gamma),$$

which is $\pi_{D,Z,\gamma}$ as required.   *q.e.d.*

According to the little-group method now $\tilde{P}_{s;2t}(F)^\wedge$ consists of the unitarily induced classes

(10.4a) $[\pi_{D,Z,\gamma,\mu}] = [\mathrm{Ind}_{\tilde{G}_{s;2t}(F) \cdot M_{D,Z} \uparrow \tilde{P}_{s;2t}(F)} (\tilde{\pi}_{D,Z,\gamma} \otimes \mu)]$

where

(10.4b) $D \in \mathrm{Sym}\, F^{s \times s}$, $Z \in F^{s \times 2t}$ as in Lemma 9.14, $[\gamma] \in (\tilde{L}_{D,Z})^{\wedge}$

and

(10.4c) $[\mu] \in \hat{M}_{D,Z}$ is extended by $\mu(D',Z',\tilde{g},\beta) = \mu(\beta)$.

Writing ε for the coverings $\tilde{P}_{s;2t}(F) \to P_{s;2t}(F)$ and $\tilde{G}_{s;2t}(F) \to G_{s;2t}(F)$
induced by $\varepsilon: \mathrm{Mp}(t;F) \to \mathrm{Sp}(t;F)$, we identify

(10.5a) $P_{s;2t}(F)^{\wedge} = \{[\pi] \in \tilde{P}_{s;2t}(F)^{\wedge}:\ \pi \text{ has the form } \bar{\pi} \cdot \varepsilon\}$

and

(10.5b) $G_{s;2t}(F)^{\wedge} = \{[\pi] \in \tilde{G}_{s;2t}(F)^{\wedge}:\ \pi \text{ has form } \bar{\pi} \cdot \varepsilon\}.$

Then

(10.5c) $P_{s;2t}(F)^{\wedge}$; $\{[\pi_{D,Z,\gamma,\mu}] \text{ from } (10.4):\ [\pi_{D,Z,\gamma}] \in G_{s;2t}(F)^{\wedge}\}.$

Two classes $[\pi_{D,Z,\gamma,\mu}] = [\pi_{D',Z',\gamma',\mu'}]$ if and only if there exists
$(g,\beta) \in \mathrm{Sp}(t;F) \times \mathrm{GL}(s;F)$ such that

(10.6a) $D' = {}^{t}\beta \cdot D\beta$ and $Z' - {}^{t}\beta \cdot Zg \in D' \cdot F^{s \times 2t}$

(10.6b) $\tilde{g}' \to \gamma(g^{-1}\tilde{g}^{t}g)$ is equivalent to γ' and

(10.6c) $\beta' \to \mu(\beta^{-1}\beta'\beta)$ is equivalent to μ'.

    With Lemma 10.1 in mind, we denote

(10.7a) $J_{d;u,2v}(\mathbb{C}) = \mathbb{C}^{d\times(u+2v)} \cdot \{O(d;\mathbb{C}) \times L^{Sp}_{u,2v}(\mathbb{C})\}$

and

(10.7b) $J_{d',d'';u,2v}(\mathbb{R}) = \mathbb{R}^{(d',d'')\times(u+2v)} \cdot \{O(d',d'') \times L^{Sp}_{u,2v}(\mathbb{R})\}$

where, using (1.3a),

(10.7c) $L^{Sp}_{u,2v}(F) = \{(x;a,b) \in L_{u,2v}(F): b \in Sp(v;F)\}.$

Schematically, J_d or $d',d'';u,2v(F)$ is given by the matrices

$$
\left[
\begin{array}{c|cc}
O(d;\mathbb{C}) \text{ or } O(d',d'') & \multicolumn{2}{c}{F^{d\times(u+2v)}} \\
\hline
 & GL(u;F) & F^{u\times2v} \\
0 & & \\
 & 0 & Sp(v;F)
\end{array}
\right] .
$$

    With this, we state and prove the principal result for maximal parabolic subgroups of symplectic and metaplectic groups.

    <u>10.8. Theorem</u>. $\tilde{P}_{s;2t}(F)^{\wedge}$ *is the disjoint union of non-empty subsets* $\tilde{S}(a,b;d)$ *where*

(*) $\begin{cases} F = \mathbb{C}: \ a,b,d \quad \text{integers} \geqslant 0, \ a + 2b + d \leqslant s \quad, \ a + b \leqslant t \\ F = \mathbb{R}: \ a,b,d^t,d'' \ \text{integers} \geqslant 0, \ a + 2b + d^t + d'' \leqslant s, \ a + b \leqslant t \end{cases}$

<u>Let</u> $(a,b;d)$ <u>satisfy</u> (*), $d = (d^t,d'')$ <u>when</u> $F = \mathbb{R}$. <u>Define</u> $D = \begin{pmatrix} D^t & 0 \\ 0 & 0 \end{pmatrix}$

in Sym $F^{s \times s}$ *by* $D' = I_d$ *if* $F = C$, $D' = \begin{pmatrix} I_{d'} & 0 \\ 0 & -I_{d''} \end{pmatrix}$ *if* $F = R$. *Choose* $Z \in F^{s \times 2t}$ *such that*

$$DZ = 0, \quad \text{rank } Z = a + 2b \ \underline{and} \ \text{rank } \mathcal{B}(Z,Z) = 2b.$$

Then $\tilde{S}(a;b;d)$ *is parameterized by*

$$\tilde{G}_{a;2(t-a-b)}(F)^{\wedge} \times J_{d;s-(d \text{ or } d' + d'')-2b,2b}(F)^{\wedge}$$

under

$$([\gamma],[\mu]) \to [\pi_{D,Z,\gamma,\mu}]$$

where

$$\tilde{L}_{D,Z} \ \textit{is identified with its isomorph} \ \tilde{G}_{a;2(t-a-b)}(F)$$

and

$$M_{D,Z} \ \textit{is identified with its isomorph} \ J_{d;s-(d \text{ or } d' + d'')-2b;2b}(F).$$

Note. $\tilde{G}_{a;2(t-a-b)}(F)^{\wedge}$ was described in Theorem 9.33. For continuity of exposition we defer the description of $J_{d;s-(d \text{ or } d' + d'')-2b;2b}(F)^{\wedge}$.

Proof. Let $[\pi_{D,Z,\gamma,\mu}] \in \tilde{P}_{s;2t}(F)^{\wedge}$. We conjugate by an element $\beta \in GL(s;F)$. This sends D to $^t\beta^{-1} \cdot D \cdot \beta^{-1}$. Thus we may assume $D = \begin{pmatrix} D' & 0 \\ 0 & 0 \end{pmatrix}$ where $D' = I_d$ if $F = C$ and $D' = \begin{pmatrix} I_{d'} & 0 \\ 0 & -I_{d''} \end{pmatrix}$ if $F = R$. Now $Z = \begin{pmatrix} 0 \\ Z' \end{pmatrix}$ with Z' in $C^{(s-d) \times 2t}$ or $R^{(s-d'-d'') \times 2t}$, so Z has rank $a + 2b \leq \min(s-(d \text{ or } d' + d''),2t)$ where $\mathcal{B}(Z,Z)$ has rank $2b$ and $a + b \leq t$. These numerical conditions are equivalent to (*).

As in the proof of Lemma 10.1, fixing D we can still vary Z by a transformation $Z' \to \beta'Z'g$ with $\beta' \in GL(s-d;F)$ and $g \in Sp(t;F)$. As

$$\mathcal{B}(\beta'Z'g,\beta'Z'g) = \beta' \cdot \mathcal{B}(Z',Z') \cdot {}^t\beta'$$

this sends

$$\mathcal{B}(Z,Z) = \begin{bmatrix} 0 & 0 \\ 0 & \mathcal{B}(Z',Z') \end{bmatrix} \to \begin{bmatrix} 0 & 0 \\ 0 & \beta' \cdot \mathcal{B}(Z',Z') \cdot {}^t\beta' \end{bmatrix}.$$

The only invariant is $b = \frac{1}{2}$ rank $\mathcal{B}(Z,Z)$.

10.9. Corollary. $P_{s;2t}(F)^\wedge$ *is the disjoint union of non-empty subsets* $S(a,b;d)$ *as in Theorem* 10.8, *with* $S(a,b;D)$ *parameterized by*

$$G_{a;2(t-a-b)}(F)^\wedge \times J_{d;s-(d \text{ or } d^t+d'')-2b;2b}(F)^\wedge$$

under

$$([\gamma], [\mu]) \leftrightarrow [\pi_{D,Z,\gamma,\mu}].$$

    Proof.   Combine (10.5) with Theorem 10.8.   *q.e.d.*

§11. Representations of the Little-Groups $J_{d;u,2v}(F)$.

We complete the determination of the irreducible unitary representations
of the maximal parabolic subgroups

$$P_{s;2(n-s)}(F) \subset Sp(n;F) \quad \text{and} \quad \tilde{P}_{s;2(n-s)}(F) \subset Mp(n;F)$$

of the symplectic and metaplectic groups, given by Theorems 8.6 and 10.8.
This is a matter of describing the unitary duals of the little-groups
$J_{d \text{ or } d',d'';u,2v}(F)$ encountered in Theorem 10.8. Here it is necessary to
look ahead to the results of §12 on orthogonal groups, and the reader may
wish to glance at that material before proceding.

Given non-negative integers d, d', d", u, v we denote

(11.1a) $I_{d;u+2v}(C) = C^{d \times (u+2v)} \cdot O(d;C)$

and, writing d for (d',d") in the real case,

(11.1b) $I_{d;u+2v}(R) = R^{d \times (u+2v)} \cdot O(d) = R^{(d',d'') \times (u+2v)} \cdot O(d',d")$.

Glancing back at (10.7) we see that $I_{d;u+2v}(F)$, is a closed normal subgroup
of $J_{d;u,v}(F)$, and that the latter is a semi-direct product

(11.2a) $J_{d;u,2v}(F) = I_{d;u+2v}(F) \cdot L_{u,2v}^{Sp}(F)$

where

(11.2b) $L_{u,2v}^{Sp}(F) = F^{u \times 2v} \cdot \{Sp(v;F) \times GL(u;F)\} \cong P_{u;2v}(F)/Sym \ F^{u \times u}$.

§11A. The Case $F = R$.

Looking back to (2.8) and Corollary 2.14 we see

(11.3) $I_{d',d'';c}(R) \cong G_{c;d',d''}(R)/\mathrm{Im}\ R^{c \times c}$.

$\mathrm{Im}\ R^{c \times c}$ is central in the unipotent radical $N_{c;d',d''}(R)$ of $G_{c;d',d''}(R)$,
and $N_{c;d',d''}(R)/\mathrm{Im}\ R^{c \times c}$ is the real vector group $R^{c \times (d',d'')}$. In the
notation (5.9) that tells us

(11.4) $I_{d',d'';c}(R)^{\wedge} = \{[\pi_{D,Z,\gamma}] \in G_{c;d',d''}(R)^{\wedge}: D = 0\}.$

Now Theorem 5.12 specializes to

    11.5. Proposition. $I_{d',d'';c}(R)^{\wedge}$ *is the disjoint union of non-empty*
subsets $S_I(k,\ell,m)$, *where* k, ℓ *and* m *are non-negative integers such*
that

(*) $k + \ell + m \leqslant c$, $k + m \leqslant d'$ *and* $\ell + m \leqslant d''$,

given as follows. Let $\mathcal{H}_{k,\ell}$ *denote the space of all real symmetric* $c \times c$
matrices H *of "signature"* $(k,\ell,c-k-\ell)$. *Then* $S_I(k,\ell,m)$ *is parameterized*
by $\mathcal{H}_{k,\ell} \times G_{m;d'-k-m;d''-\ell-m}(R)^{\wedge}$ *under*

        $(H,[\gamma]) \leftrightarrow [\pi_{0,Z,\gamma}]$ *in the notation* (11.4) *and* (5.9)

where $Z \in R^{c \times (d',d'')}$ *with* $\mathcal{H}(Z,Z) = H$ *and* rank $Z = k + \ell + m$.
    Let $c = u + 2v$. The $L_{u,2v}^{Sp}(R)$-stabilizer of $[\pi_{0,Z,\gamma}] \in I_{d',d'';c}(R)^{\wedge}$
is obtained by specializing Lemma 6.1 to the case $D = 0$, so it is

(11.6 a) $M_Z = \{\beta \in L^{Sp}_{u,2v}(R) \subset GL(c;R): {}^t\beta \cdot Z \in Z \cdot O(d',d'')\}$.

Restricting the extensions provided by Lemma 6.6, we have

(11.6b)   every $[\pi_{0,Z,\gamma}] \in I_{d',d'';c}(R)^{\wedge}$ extends to $I_{d',d'';c}(R) \cdot M_Z$.

Now, by the little-group method,

   11.7. Theorem. $J_{d',d'';u,2v}(R)^{\wedge}$ _is the disjoint union of non-empty sub-_
sets $S_J(m,\{H\})$ _where_ k , ℓ _and_ m _are non-negative integers such that_

$$k + \ell + m \leqslant u + 2v, \quad k + m \leqslant d' \quad \underline{and} \quad \ell + m \leqslant d''$$

and where $\{H\}$ _is an_ $L^{Sp}_{u,2v}(R)$-_equivalence class of_ $(u + 2v) \times (u + 2v)$
symmetric real matrices H _of "signature"_ $(k,\ell,u + 2v-k-\ell)$. $S_J(m,\{H\})$
is parameterized by $G_{m;d'-k-m,d''-\ell-m}(R)^{\wedge} \times \hat{M}_Z$ _under_

$$([\gamma],[\mu]) \leftrightarrow [\mathrm{Ind}_{I_{d',d'';u+2v}(R) \cdot M_Z \uparrow J_{d',d'';u,2v}(R)}(\tilde{\pi}_{0,Z,\gamma} \otimes \mu)]$$

where $Z \in R^{c \times (d',d'')}$ _with_ $\mathcal{H}(Z,Z) = H$ _and_ rank $Z = k + \ell + m$.

   The determination of $J_{d',d'';u,2v}(R)^{\wedge}$ now is reduced to that of the
$\hat{M}_Z$. We are about to see that the M_Z are groups of similar type composed
of smaller matrices, so we may assume $\hat{M}_Z$ known by recursion of the size
of the matrices.

   We complete the discussion of $J_{d',d'';u,2v}(R)$ as mentioned above. Let
$c = u + 2v$ and note that we have

(11.8a)   an antisymmetric bilinear form α of rank $2v$ on R^c

and

(11.8b)   a symmetric bilinear form β of "signature" $(k,\ell,c-k-\ell)$ on R^c

such that

(11.8c) $M_Z = \{g \in GL(c;R): \ g \ \text{preserves both} \ \alpha \ \text{and} \ \beta\}$.

In effect, $L^{Sp}_{u,2v}(R) = \{g \in GL(c;R): \ g \ \text{preserves} \ \alpha\}$, and the condition
(11.6a) that ${}^t g \cdot Z \in Z \cdot O(d',d'')$ just says that g preserves the symmetric
bilinear form β with matrix $\mathcal{H}(Z,Z)$. Now split $R^c = U \oplus V \oplus W$ where

$U \oplus V$ is the null space of β , so $\beta|_{W \times W}$ is nondegenerate,

U is the null space of $\alpha|_{(U+V) \times (U+V)}$, so $\alpha|_{V \times V}$ is nondegenerate,

W is the orthogonal to V relative to α .

The elements of M_Z preserve $U \oplus V$, preserve U , and send W into $U \oplus W$.
Thus $g \in M_Z$ is given schematically by

$$\begin{bmatrix} g_{11} & g_{12} & g_{13} \\ 0 & g_{22} & 0 \\ 0 & 0 & g_{33} \end{bmatrix} \in \begin{bmatrix} GL(U) & U \otimes V^* & U \otimes W^* \\ 0 & Sp(\alpha|_{V \times V}) & 0 \\ 0 & 0 & O(\beta|_{W \times W}) \end{bmatrix}$$

subject only to the restriction, relating g_{13} and g_{33} , that
$\alpha(gw,gw') = \alpha(w,w')$ for all $w, w' \in W$. Set $n_1 = \dim U$, $n_2 = \dim V$ and
$n_3 = \dim W$. Now $M_Z \cong I''_{n_1,n_2;n_3}(R) \cdot L^{Sp}_{n_1,n_2}(R)$ where

$L^{Sp}_{n_1,n_2}(R) = \{(x;a,b) \in L_{n_1,n_2}(R): \ b \in Sp(\frac{1}{2}n_2;R)\}$ as in (10.7c) and (11.2b),
and where $I''_{n_1,n_2;n_3}(R)$ is the subgroup $(g|_{U+V} = \text{identity})$ of M_Z , which
sits naturally in $R^{n_1 \times n_3} \cdot O(\beta|_{W \times W}) = R^{n_1 \times (k,\ell)} \cdot O(k,\ell)$ under

$$\begin{bmatrix} I & 0 & g_{13} \\ 0 & I & 0 \\ 0 & 0 & g_{33} \end{bmatrix} \rightarrow \begin{bmatrix} I & g_{13} \\ 0 & g_{33} \end{bmatrix} \rightarrow (g_{13}\, g_{33}^{-1}\,,\, g_{33}).$$

Given specific α and β, we now know enough to write out the elements of $\hat{M}_Z$, and so $J_{d_1,d_2;u,2v}(R)\hat{}$ is determined.

$$\S 11B. \quad \text{The Case} \quad F = C.$$

Looking ahead to (12.5) and Theorem 12.6 we see that

(11.9) $\qquad I_{d;c}(C) \cong G_{c;d',d''}(R)_C / \text{Skew } C^{c \times c} \quad$ whenever $\quad d = d' + d''.$

Skew $C^{c \times c}$ is central in the unipotent radical $N_{c;d',d''}(R)_C$ of $G_{c;d',d''}(R)_C$, and $N_{c;d',d''}(R)_C / \text{Skew } C^{c \times c}$ is a vector group $C^{c \times (d',d'')}$. In the notation (12.27), that gives

(11.10) $\qquad I_{d;c}(C)\hat{} = \{[\pi_{D,Z,\gamma}] \in G_{c;d',d''}(R)\hat{_C} : D = 0\}\ .$

Now Theorem 12.28 specializes to

11.11. Proposition. $I_{d;c}(C)\hat{}$ *is the disjoint union of non-empty subsets* $S_I(a,b)$ *where* a *and* b *are non-negative integers such that* $a + b \leq c$ *and* $2a + b \leq d$, *given as follows. Decompose*

$$d = d' + d'' \text{ and } b = b' + b'' \text{ with } b^{(i)}, d^{(i)} \geq 0 \text{ and } a + b^{(i)} \leq d^{(i)}.$$

Let $\mathcal{S}_b$ *denote the space of all* c × c *symmetric complex matrices* S *of rank* b. *Then* $S_I(a,b)$ *is parameterized by* $\mathcal{S}_b \times G_{a;d'-b'-a,d''-b''-a}(R)\hat{_C}$ *under*

$$(S, [\gamma]) \leftrightarrow [\pi_{0,Z,\gamma}] \quad \textit{in the notation (11.10) and (12.27)}$$

where $Z \in C^{c \times (d',d'')}$ *with* $\mathcal{B}(Z,Z) = S$ *and* rank $Z = a+b$.

Let $c = u + 2v$. The $L^{Sp}_{u,2v}(C)$-stabilizer of $[\pi_{0,Z,\gamma}] \in I_{d;c}(C)^{\wedge}$ is obtained by specializing Lemma 12.29 to the case $D = 0$, so it is

$$(11.12a) \qquad M_Z = \{\beta \in GL(c;C) : {}^t\beta \cdot Z \in Z \cdot O(d;C)\}.$$

Restricting the result of Lemma 12.31,

$$(11.12b) \qquad \text{every } [\pi_{0,Z,\gamma}] \in I_{d;c}(C)^{\wedge} \text{ extends to } I_{d;c}(C) \cdot M_Z.$$

Now, using Mackey's little-group method,

11.13. Theorem. $J_{d;u,2v}(C)^{\wedge}$ *is the disjoint union of non-empty subsets* $S_J(a;\{S\})$ *where* a *and* b *are non-negative integers such that*

$$a + b \le u + 2v \quad \textit{and} \quad 2a + b \le d$$

and where $\{S\}$ *is an* $L_{u,2v}(C)$*-equivalence class of* $(u + 2v) \times (u + 2v)$ *symmetric complex matrices* S *of rank* b. *Choose splittings*

$$d = d' + d'' \quad \textit{and} \quad b = b' + b'' \textit{ with } b^{(i)}, d^{(i)} \ge 0 \textit{ and } a + b^{(i)} \le d^{(i)}.$$

Then $S_J(a:\{S\})$ *is parameterized by* $G_{a;d'-b'-a,d''-b''-a}(R)^{\wedge}_C \times \hat{M}_Z$ *under*

$$([\gamma],[\mu]) \leftrightarrow [\text{Ind}_{I_{d;u+2v}(C) \cdot M_Z} \uparrow J_{d;u,v}(C)(\tilde{\pi}_{0,Z,\gamma} \otimes \mu)]$$

where $Z \in C^{c \times (d',d'')}$ *with* $\mathcal{B}(Z,Z) = S$ *and* rank $Z = a+b$.

We have reduced the investigation of $J_{d;u,2v}(C)^\wedge$ to the determination of the $\hat{M}_Z$. Now we will show that the M_Z are groups of similar type, based on matrices of smaller size, so that we may take $\hat{M}_Z$ to be known by recursion on the size of the matrices.

To complete the discussion of $J_{d;u,2v}(C)$ in the manner just described, let $c = u + 2v$ and note that we have

(11.14a)   an antisymmetric bilinear form α of rank $2v$ on C^c

and

(11.14b)   a symmetric bilinear form β of rank b on C^c

such that

(11.14c) $M_Z = \{g \in GL(c;C): \ g \ \text{preserves both} \ \alpha \ \text{and} \ \beta\}$.

For $L^{Sp}_{u,2v}(C) = \{g \in GL(c;C): \ g \ \text{preserves} \ \alpha\}$ and the condition (11.12a) that $^t g \cdot Z \in Z \cdot O(d;C)$ just says that g preserves the symmetric form β with matrix $\mathcal{B}(Z,Z)$. We split $C^c = U \oplus V \oplus W$ where

   $U \oplus V$ is the null space of β, so $\beta|_{W \times W}$ is nondegenerate,
   U is the null space of $\alpha|_{(U+V) \times (U+V)}$, so $\alpha|_{V \times V}$ is nondegenerate,
   W is orthogonal to W under α.

Every element of M_Z preserves $U \oplus V$, preserves U, and maps W into $U \oplus W$. So $g \in M_Z$ is schematically by

$$\begin{bmatrix} g_{11} & g_{12} & g_{13} \\ 0 & g_{22} & 0 \\ 0 & 0 & g_{33} \end{bmatrix} \in \begin{bmatrix} GL(U) & U \otimes V^* & U \otimes W^* \\ 0 & Sp(\alpha|_{V \times V}) & 0 \\ 0 & 0 & O(\beta|_{W \times W}) \end{bmatrix}$$

with g_{13} related to g_{33} by: $\alpha(gw, gw') = \alpha(w, w')$ for all $w, w' \in W$.

Set $n_1 = \dim U$, $n_2 = \dim V$ and $n_3 = \dim W$. Now

$M_Z \cong I''_{n_1, n_2; n_3}(\mathbb{C}) \cdot L^{Sp}_{n_1, n_2}(\mathbb{C})$ where

$L^{Sp}_{n_1, n_2}(\mathbb{C}) = \{(\chi; \sigma, \tau) \in L_{n_1, n_2}(\mathbb{C}): \tau \in Sp(\frac{1}{2}n_2; \mathbb{C})\}$ as in (10.7c) and (11.2b).

Here $I''_{n_1, n_2; n_3}(\mathbb{C})$ is the subgroup $(g|_{U+V} = \text{identity})$ of M_Z; it sits

naturally in $\mathbb{C}^{n_1 \times n_3} \cdot O(n_3; \mathbb{C}) = \mathbb{C}^{n_1 \times (b', b'')} \cdot O(b', b''; \mathbb{C})$ via

$$\begin{bmatrix} I & 0 & g_{13} \\ 0 & I & 0 \\ 0 & 0 & g_{33} \end{bmatrix} \rightarrow \begin{bmatrix} I & g_{13} \\ 0 & g_{33} \end{bmatrix} \rightarrow (g_{13}\, g_{33}^{-1},\, g_{33}).$$

Given specific α and β, this tells us enough to write out $\hat{M}_Z$. So, finally, $J_{d; u, 2v}(\mathbb{C})^{\wedge}$ is determined.

<u>Part IV. Orthogonal Groups</u>

<u>§12. Parabolic Subgroups of Complex Orthogonal Groups.</u>

In this section we write out the structure and representation theory
for the maximal parabolic subgroups of the complex orthogonal groups $O(n;C)$.
Essentially this is a matter of complexifying the results of §§2-7 for an
indefinite orthogonal group $O(p,q)$, $p + q = n$.

If σ is a nondegenerate symmetric bilinear form on C^n then the
corresponding <u>complex orthogonal group</u> is $\{g \in GL(n;C): \sigma(gx,gy) = \sigma(x,y),$
all $x,y \in C^n\}$. Any two choices of σ are $GL(n;C)$-equivalent, so the
corresponding complex orthogonal groups are conjugate in $GL(n;C)$, hence
isomorphic. In this section we will use

$$(12.1a) \qquad \sigma(x,y) = \sum_{1 \leqslant \ell \leqslant m} x^\ell y^\ell - \sum_{m < \ell \leqslant n} x^\ell y^\ell , \quad n = 2m \text{ or } 2m + 1,$$

and we will denote the complex orthogonal group by the usual

$$(12.1b) \qquad O(n;C) = \{g \in GL(n;C): \sigma(gx,gy) = \sigma(x,y), \text{ all } x, y \in C^n\}.$$

In some later sections it will turn out to be more convenient to use
$\sum_{1 \leqslant \ell \leqslant m} (x^\ell y^{\ell+m} + x^{\ell+m} y^\ell)$ to specify $O(2m;C)\}$.

A linear subspace $E \subset C^n$ is <u>totally isotropic</u> if $\sigma(E,E) = 0$. The
<u>parabolic subgroups</u> of $O(n;C)$ are the

$$(12.2a) \qquad P_{E_1,\ldots,E_k} = \{g \in O(n;C): gE_\ell = E_\ell \text{ for } 1 \leqslant \ell \leqslant k\}$$

where $0 \neq E_1 \subsetneqq \ldots \subsetneqq E_k$ is a sequence of totally isotropic subspaces. If
$0 \neq E_1' \subsetneqq \ldots \subsetneqq E_k'$ is another such sequence, and if $\dim E_\ell = \dim E_\ell'$ for
$1 \leqslant \ell \leqslant k$, then some $g \in O(n;C)$ satisfies $gE_\ell' = E_\ell$ for $1 \leqslant \ell \leqslant k$, and

so $g \cdot P_{E_1', \ldots, E_k'} \cdot g^{-1} = P_{E_1, \ldots, E_k}$. Thus the dimension sequence $\{\dim E_\ell\}$ determines the conjugacy class of $P_{E_1, \ldots, E_k}$ in $O(n;C)$. In particular $P_{E_1, \ldots, E_k}$ is conjugate to $P_{E_1', \ldots, E_k'}$ where E_ℓ' is spanned by $\{e_1 + e_{m+1}, \ldots, e_{s_\ell} + e_{m+s_\ell}\}$, $s_\ell = \dim E_\ell$.

Now the maximal parabolic subgroups of $O(n;C)$ are the

(12.2b) $P_E = \{g \in O(n;C): gE = E\}$, E nonzero totally isotropic in C^n.

There are $m = [n/2]$ conjugacy classes of them as $\dim E = 1, 2, \ldots, m$.

Define a bilinear map $\mathcal{B}: C^{s \times (t,u)} \times C^{s \times (t,u)} \to C^{s \times s}$ by

(12.3a) $\mathcal{B}((A_0, B_0), (A,B)) = A_0 \cdot {}^t A - B_0 \cdot {}^t B$ where $A, A_0 \in C^{s \times t}$ and
          $B, B_0 \in C^{s \times u}$.

Decompose $C^{s \times s} = \text{Sym } C^{s \times s} \oplus \text{Skew } C^{s \times s}$ under the projections

(12.3b) $\text{Sym}(D) = \frac{1}{2}(D + {}^t D)$ and $\text{Skew}(D) = \frac{1}{2}(D - {}^t D)$.

Then we have nilpotent complex Lie groups

(12.4a) $N_{s;t,u}(R)_C = \text{Skew } C^{s \times s} + C^{s \times (t,u)}$ with group law

(12.4b) $(D_0, Z_0)(D, Z) = (D_0 + D + \text{Skew } \mathcal{B}(Z_0, Z), Z_0 + Z)$

and with Lie algebra given by

(12.4c) $\mathfrak{n}_{s;t,u}(R)_C = \text{Skew } C^{s \times s} + C^{s \times (t,u)}$ where

          $[(D_0, Z_0), (D, Z)] = (2 \text{ Skew } \mathcal{B}(Z_0, Z), 0)$.

The group $N_{s;t,u}(R)_C$ of (12.4) is the complexification of the nilpotent real Lie group $N_{s;t,u}(R)$ of (2.4). In particular it is abelian just when $s \leqslant 1$ or $t + u = 0$, and if non-abelian then it is 2-step nilpotent with center Skew $C^{s \times s}$.

Let $\sigma_{t,u}$ denote the symmetric bilinear form on C^{t+u} given by

$$\sigma_{t,u}(x,y) = \sum_{1 \leqslant \ell \leqslant t} x^\ell y^\ell - \sum_{1 \leqslant \ell \leqslant u} x^{t+\ell} y^{t+\ell},$$ and let $O(t,u;C)$ denote its ortho-

gonal group. Of course $O(t,u;C) \cong O(t + u;C)$, but we are going to have to make a distinction.

$O(t,u;C) \times GL(s;C)$ acts by automorphisms on $N_{s;t,u}(R)_C$ by

(12.5a) $(g,\gamma): (D,W) \to (\gamma D \cdot {}^t\gamma, \gamma Z \cdot {}^t g)$

and we form the semidirect product groups

(12.5b) $G_{s;t,u}(R)_C = N_{s;t,u}(R)_C \cdot O(t,u;C)$ and

(12.5c) $P_{s;t,u}(R)_C = N_{s;t,u}(R)_C \cdot \{O(t,u;C) \times GL(s;C)\}.$

The complex Lie groups (12.5) are the respective complexifications of the real Lie groups $G_{s;t,u}(R)$ of (2.8a) and $P_{s;t,u}(R)$ of (2.9a). The maximal unimodular subgroup of $P_{s;t,u}(R)_C$ is the real group

$P_{s;t,u}(R)'_C = N_{s;t,u}(R)_C \cdot \{O(t,u;C) \times GL'(s;C)\}.$

12.6. Theorem. *Let* E *be a totally isotropic subspace of dimension* $s > 0$ *in* C^n, *and let* P_E *be the corresponding maximal parabolic subgroup of* $O(n;C)$. *There is a biholomorphic Lie group isomorphism*

(12.7a) $\varphi: P_{s;m-s,n-m-s}(R)_C \to P_E$

that carries $N_{s;m-s,n-m-s}(R)_C$ *to the unipotent radical and*

$O(m-s,n-m-s;C) \times GL(s;C)$ _to a reductive complement, and has restriction_

(12.7b) $\varphi: G_{s;m-s,n-m-s}(R)_C \cong \{g \in O(n;C): g|_E = \underline{identity}\}.$

Theorem 12.6 follows from the case $F = R$ of Theorem 2.10: complexify everything in its proof. Thus the Langlands decomposition $P_E = MAN$ is reflected by

$$N = \varphi \cdot N_{s;m-s,n-m-s}(R)_C, \quad A = \varphi \cdot R^+ \quad \text{and} \quad M = \varphi \cdot \{O(m-s,n-m-s;C) \times GL'(s;C)\}.$$

Further,

12.8. Corollary. _A maximal parabolic subgroup_ P_E _of_ $O(n;C)$ _has abelian unipotent radical if, and only if, either_ $\dim E = 1$ _or_ $n = 2 \cdot \dim E$.

12.9. Corollary. _A maximal parabolic subgroup_ P_E _of_ $O(n;C)$ _is cuspidal if, and only if, both_ $\dim E = 1$ _and_ $2 \leqslant n \leqslant 3$.

To discuss the nilradical we denote

(12.10a) $N = N_{s;t,u}(R)_C$, simply connected nilpotent group;

(12.10b) $n = n_{s;t,u}(R)_C = \text{Skew } C^{s \times s} + C^{s \times (t,u)}$, its Lie algebra;

(12.10c) n^*: the real linear dual space of n; and

(12.10d) Ad^*: the (co-adjoint) representation of N on n^*.

We have a nondegenerate real inner product on n given by

(12.11a) $\langle (D_0, Z_0), (D, Z) \rangle = \text{trace Re } D_0 \cdot {}^t D + \text{trace Re } \mathscr{B}(Z_0, Z)$

where Re denotes real part. That identifies

(12.11b) $n^* = \{f_{D,Z}: (D,Z) \in n\}$ where $f_{D,Z}(D_0,Z_0) = \langle(D_0,Z_0),(D,Z)\rangle$.

As usual we use f_D for $f_{D,0}$ and g_Z for $f_{0,Z}$. Now, as in Propositions 4.3 and 9.3,

    12.12. Proposition. *Let* $D \in$ Skew $C^{s \times s}$. *Then we have a direct sum decomposition*

(12.13a) $C^{s \times (t,u)} = \{Z \in C^{s \times (t,u)}: DZ = 0\} \oplus C_D^{s \times (t,u)}$

with the provision: if the matrix D *is diagonalizable then* $C_D^{s \times (t,u)} = D \cdot C^{s \times (t,u)}$. *If* $f = f_{D,Z} \in n^*$, *then the isotropy algebra* $n_f = \{x \in n: f[x,n] = 0\}$ *is*

(12.13b) $n_f =$ Skew $C^{s \times s} + \{Z^t \in C^{s \times (t,u)}: DZ' = 0\}$

and the co-adjoint orbit is the affine subspace

12.13c) $Ad^*(N) \cdot f = f + \{g_{DZ'}: Z' \in C_D^{s \times (t,u)}\}$.

N has square integrable representations just when n^* has an element $f = f_{D,Z}$ such that n_f is the center of n. That happens only when N is abelian or D is nonsingular. In other words,

    12.14. Corollary. $N_{s;t,u}(R)_C$ *has square integrable representations if and only if* (i) $s \leqslant 1$, *or* (ii) $t + u = 0$, *or* (iii) s *is even.*

As usual, for fixed $f = f_{D,Z}$ we denote

(12.15a) $\mathfrak{d}_f = \{D^t \in$ Skew $C^{s \times s}: f(D^t) = 0\}$, central ideal in n;

(12.15b) $\mathfrak{m}_f = \mathfrak{n}/\mathfrak{d}_f$ quotient algebra and p: $\mathfrak{n} \to \mathfrak{m}_f$ projection;

(12.15c) $\mathfrak{h}_f = p(\text{Skew } C^{s \times s} + C_D^{s \times (t,u)})$ with $C_D^{s \times (t,u)}$ as in (12.13a);

and

(12.15d) $\alpha_f = p(\mathfrak{d}_f + \{Z' \in C^{s \times (t,u)} : DZ' = 0\})$.

As in §4, $\mathfrak{m}_f = \mathfrak{h}_f \oplus \alpha_f$, direct sum of ideals. Setting

(12.16a) M_f: the simply connected nilpotent Lie group for $\mathfrak{m}_f$, and

(12.16b) H_f and A_f: the analytic subgroups for $\mathfrak{h}_f$ and α_f,

we have

(12.16c) $M_f = H_f \times A_f$.

A_f is a vector group, and the possibilities for H_f are

(12.17a) H_f is a real Heisenberg group of dimension ≥ 3 (D $\neq$ 0,
            t + u > 0),

(12.17b) H_f is a 1-dimensional real vector group (D $\neq$ 0, t + u = 0),

(12.17c) H_f is a the trivial group (D = 0).

According to the Kirillov Theory, the class $[\pi_f] \in \hat{N}$ corresponding to the co-adjoint orbit of f is

(12.18a) $[\pi_f] = [\bar{\pi}_f \cdot p]$ where $[\bar{\pi}_f] \in \hat{M}_f$ is the class for $\bar{f}$, f = $\bar{f} \cdot p$,
and (12.16c) gives us

(12.18b) $[\bar{\pi}_f] = [\eta_f \otimes \alpha_f]$ where $[\eta_f] \in \hat{H}_f$ and $\alpha_f \in \hat{A}_f$.

Here $\alpha_f = e^{ig_Z} : \exp(\mathfrak{d}_f + Z') \to e^{ig_Z(Z')} = e^{i \text{ trace Re } \mathcal{B}(Z, Z')}$ and η_f acts on $p(\exp \text{Skew } C^{s\times s})$ by the unitary character e^{if_D}: $\exp(D' + \dot{a}_f) \to e^{if_D(D')} = e^{-i \text{ trace Re } DD'}$. According to the three cases of (12.17),

(12.19a) H_f is Heisenberg and $\eta_f \in \hat{H}_f$ has central character $e^{if_D} \neq 1$,

(12.19b) $H_f \cong R^1$ and η_f is the unitary character $e^{if_D} \neq 1$,

(12.19c) H_f and η_f are trivial.

We summarize, describing $\hat{N}$:

    12.20. Theorem. *Let* $N = N_{s;t,u}(R)_C$. *Then* $\hat{N}$ *consists of the classes* $[\pi_{D,Z}] = [\pi_{f_{D,Z}}]$ *given as follows.*

(12.21a) $D \in \text{Skew } C^{s\times s}$ *and* $Z \in F^{s\times(t,u)}$ *modulo* $D \cdot F^{s\times(t,u)}$,

(12.21b) $p: N \to M_f = H_f \times A_f$ *as in* (12.15) *and* (12.16) *with* $f = f_{D,Z}$,

(12.21c) α_f *is the unitary character* e^{ig_Z} *on* A_f,

(12.21d) $[\eta_f] \in \hat{H}_f$ *is the class given by* (12.17) *and* (12.19), *and*

(12.21e) $[\pi_{D,Z}] = [(\eta_f \otimes \alpha_f) \cdot p] \in \hat{N}$.

Further,

    (i) $[\pi_{D,Z}] = [\pi_{D',Z'}]$ *if and only if* $D = D'$ *and*
        $Z - Z' \in D \cdot C^{s\times(t,u)}$

    (ii) *the central character of* $[\pi_{D,Z}]$ *restricts to* e^{if_D} *on*
        Skew $C^{s\times s}$,

    (iii) *if* $D \neq 0$ *and* $t + u > 0$ *then* $[\pi_{D,Z}]$ *is infinite-dimensional,*

    (iv) *if* $D = 0$ *or* $t + u = 0$ *then* $[\pi_{D,Z}]$ *is a unitary character.*

Finally, the Plancherel measure of $\hat{N}$ *is concentrated on*
$\{[\pi_{D,Z}]:$ *the matrix* $D \in \text{Skew } \mathbb{C}^{s \times s}$ *has rank* $2[s/2]\}$.

The proof of Lemma 5.2 gives us

    <u>12.22. Lemma.</u> $O(t,u;\mathbb{C}) \times GL(s;\mathbb{C})$ *acts on* $\mathfrak{n}^*$ *by*
$$(g,\gamma)^{-1} f_{D,Z} = f_{t_{\gamma \cdot D}, t_{\gamma Z g}} \quad .$$

    $D \in \text{Skew } \mathbb{C}^{s \times s}$ is $GL(s;\mathbb{C})$-equivalent to $\begin{bmatrix} D' & 0 \\ 0 & 0 \end{bmatrix}$ where
$D' = \begin{bmatrix} 0 & I \\ -I & 0 \end{bmatrix} \in \text{Skew } \mathbb{C}^{2r \times 2r}$, $2r = \text{rank } D$. Now the argument of Lemma 9.14
gives us

    <u>12.23. Lemma.</u> *Let* $L_{D,Z}$ *denote the* $O(t,u;\mathbb{C})$-*stabilizer of* $[\pi_{D,Z}]$.
Then

(12.24a) $\qquad\qquad L_{D,Z} = \{g \in O(t,u;\mathbb{C}): Zg - Z \in D \cdot \mathbb{C}^{s \times (t,u)}\}.$

Choose $\gamma \in GL(s;\mathbb{C})$ *with* $t_{\gamma} \cdot D\gamma$ *semisimple.* *Without changing* $[\pi_{D,Z}]$ *we*
add an element of $D \cdot \mathbb{C}^{s \times (t,u)}$ *to* Z *so that* $(t_{\gamma} \cdot D\gamma)(t_{\gamma} \cdot Z) = 0$. *Assuming*
that normalization of Z, *let* S_Z *denote the subspace of* $\mathbb{C}^{t,u}$ *spanned*
by the columns of t_Z. *Then*

(12.24b) $\qquad L_{D,Z} = \{g \in O(t,u;\mathbb{C}): Zg = Z\}$
$$\qquad\qquad\qquad = \{g \in O(t,u;\mathbb{C}): gx = x \text{ } \text{for all} \text{ } x \in S_Z\}.$$

    Theorem 12.6 now combines with the orthocomplementation argument used
in the proof of Proposition 5.6 to give us the structure of $L_{D,Z}$:

    <u>12.25. Proposition.</u> *Let* S *be a subspace of* $\mathbb{C}^{t+u}$. *Using* $\sigma_{t,u}$,
decompose $S = (S \cap S^{\perp}) + \mathbb{C}^{a,b}$, $a + c \leqslant t$ *and* $b + u \leqslant s$ *where*

$c = \dim S \cap S^{\perp}$. *Then* $\{g \in O(t,u;\mathbb{C}): g|_S = \mathit{identity}\} \cong G_{c;t-c-a,u-c-b}(\mathbb{R})_{\mathbb{C}}$.

We extend $[\pi_{D,Z}]$ to $N \cdot L_{D,Z}$ by a reduction to Proposition 5.8.

12.26. Proposition. *Let* $[\pi_{D,Z}] \in \{N_{s;t,u}(\mathbb{R})_{\mathbb{C}}\}^{\wedge}$ *as in Theorem* 12.20. *Then* $\pi_{D,Z}$ *extends to a unitary representation* $\tilde{\pi}_{D,Z}$ *of* $N_{s;t,u}(\mathbb{R})_{\mathbb{C}} \cdot L_{D,Z}$ *on the same Hilbert space.*

Proof. Let $f = f_{D,Z}$. This is a matter of extending η_f from H_f to $H_f \cdot L_{D,Z}$. For that, we define an isomorphism $\beta: H_f \to N_{2;r(t+u),r(t+u)}(\mathbb{R})$, $2r = \operatorname{rank} D$, as follows. Replace (D,Z) by $({}^t\gamma \cdot D\gamma, {}^t\gamma \cdot Z)$ so that

$$D = \begin{bmatrix} D' & 0 \\ 0 & 0 \end{bmatrix} \text{ with } D' = \begin{bmatrix} 0 & I \\ -I & 0 \end{bmatrix} \text{ in } r \times r \text{ blocks, and then replace } Z$$

within $Z + D \cdot \mathbb{C}^{s \times (t,u)}$ so that $DZ = 0$. That done, $p: N \to M_f$ induces an isomorphism of the real subgroup

$$H_f'' = \{(D'',Z'') \in N: D'' \in D \cdot \mathbb{R} \text{ and } Z'' \in D \cdot \mathbb{C}^{s \times (t,u)}\}$$

onto H_f. Now define

$$\alpha: H_f'' \to N_{2;r(t+u),r(t+u)}(\mathbb{R}), \quad \text{Lie group isomorphism,}$$

by

$$\alpha(kD,Z'') = \left(\begin{bmatrix} 0 & k \\ -k & 0 \end{bmatrix}, \begin{bmatrix} \operatorname{Re}A^1, & \operatorname{Im}B^1, \ldots, \operatorname{Re}A^r, & \operatorname{Im}B^r; & \operatorname{Re}B^1, \operatorname{Im}A^{\perp}, \ldots, \operatorname{Re}B^r, & \operatorname{Im}A^r \\ \operatorname{Re}C^1, & \operatorname{Im}D^1, \ldots, \operatorname{Re}C^r, & \operatorname{Im}D^r; & \operatorname{Re}D^1, \operatorname{Im}C^1, \ldots, \operatorname{Re}D^r, & \operatorname{Im}C^r \end{bmatrix} \right)$$

where

$$Z'' = \begin{bmatrix} A & B \\ C & D \\ 0 & 0 \end{bmatrix}; \text{ and } A = \begin{bmatrix} A^1 \\ \vdots \\ A^r \end{bmatrix}, \quad B = \begin{bmatrix} B^1 \\ \vdots \\ B^r \end{bmatrix}, \quad C = \begin{bmatrix} C^1 \\ \vdots \\ C^r \end{bmatrix} \text{ and } D = \begin{bmatrix} D^1 \\ \vdots \\ D^r \end{bmatrix} \text{ are}$$

$r \times r$.

Then β is defined to be the composition of $p^{-1}\colon H_f \to H_f''$ with

$\alpha\colon H_f''\colon \to N_{2;r(t+u),r(t+u)}(R)$. Now the point is that β carries (the action

of) the group $L_{D,Z}$ of automorphisms of H_f over to (the action of) a sub-

group of the group $O(r(t+u),r(t+u))$ of automorphisms of

$N_{2;r(t+u),r(t+u)}(R)$. That image $\tilde{\beta}(L_{D,Z})$ stabilizes the irreducible unitary

equivalence class $[\eta] \in N_{2;r(t+u),r(t+u)}(R)^{\widehat{}}$ such that $[\eta_f] = [\eta\cdot\beta]$, so

Proposition 5.8 tells us that η extends to a unitary representation η' of

$N_{2;r(t+u),r(t+u)}(R)\cdot\tilde{\beta}(L_{D,Z})$ on the same Hilbert space. Now

$\eta_f'(n,g) = \eta'(\beta n,\tilde{\beta}g)$ is a unitary representation of $H_f\cdot L_{D,Z}$, extending η_f,

on the same space.

$$q.e.d.$$

The Mackey little-group method now tells us that $G_{s;t,u}(R)^{\widehat{}}_C$ consists

of the unitarily induced classes

(12.27a) $[\pi_{D,Z,\gamma}] = [\text{Ind}_{N_{s;t,u}(R)_C\cdot L_{D,Z}\uparrow G_{s;t,u}(R)_C}(\tilde{\pi}_{D,Z}\otimes\gamma)]$

where

(12.27b) $[\gamma] \in \hat{L}_{D,Z}$ extended as usual by $\gamma(D'.Z';g) = \gamma(g)$.

Assuming the normalization of Lemma 12.23 with D and D' semisimpli-

fied by the same element of $GL(s;C)$, we see that $[\pi_{D,Z,\gamma}] = [\pi_{D',Z',\gamma'}]$

just when $O(t,u;C)$ has an element g such that

(12.27c) $D = D'$, $Z' = Zg^{-1}$, and $g' \to \gamma(gg'g^{-1})$ is equivalent to γ'.

Observing

$$Z' \in Z\cdot O(t,u;C) \iff \text{rank } Z' = \text{rank } Z \text{ and } \mathcal{B}(Z',Z') = \mathcal{B}(Z,Z)$$

we reformulate our discussion as follows.

12.28. Theorem. $G_{s;t,u}(R)\widehat{}_C$ *is the disjoint union of non-empty subsets*
S(a+b,c;D), $D \in$ Skew $C^{s \times s}$ *and* a,b,c *non-negative integers such that*

(*) $a + b + c + \text{rank } D \leqslant s, \quad a + c \leqslant t \quad \underline{and} \quad b + c \leqslant u,$

as follows. Given D, *choose* $\beta \in$ GL(s;C) *with* $^t\beta \cdot D\beta$ *semisimple, and
let* $\mathscr{S}_{a+b,D}$ *denote the space of all* s×s *symmetric complex matrices* S *of
rank* a + b *such that* $(^t\beta \cdot D\beta)(^t\beta \cdot S\beta) = 0.$ *Then* S(a+b,c;D) *is parameterized by*

$$\mathscr{S}_{a+b,D} \times G_{c;t-c-a,u-c-b}(R)\widehat{}_C$$

under

$$(S,[\gamma]) \leftrightarrow [\pi_{D,Z,\gamma}] \quad (12.27)$$

where

$$Z \in C^{s \times (t,u)} \quad \underline{with} \quad {}^t\beta D\beta \cdot {}^t\beta Z = 0, \quad \mathscr{B}(Z,Z) = S \quad \underline{and} \quad \text{rank } Z = a+b+c,$$

and where

$L_{D,Z}$ *is identified with its isomorph* $G_{c;t-c-a,u-c-b}(R)\widehat{}_C.$

Now we start the passage from $G_{s;t,u}(R)\widehat{}_C$ to $P_{s;t,u}(R)\widehat{}_C$.

12.29. Lemma. *The* GL(s;C)-*stabilizer of a class*
$[\pi_{D,Z,\gamma}] \in G_{s;t,u}(R)\widehat{}_C$ *defined in* (12.27) *is*

$M_{D,Z} = \{\beta \in GL(s;C):\ {}^t\beta\cdot D\beta = D\ \underline{and}\ {}^t\beta\cdot Z{-}Zg \in D\cdot C^{s\times(t,u)}\ \underline{where}\ g \in O(t,u;C)\}.$

D _has even rank, say_ $2r$. _Let_ $a + b = \text{rank}\ \mathcal{B}(Z,Z)$ _after the normaliza-_
tion of Lemma 12.23 . _Then_

(12.30a) $M_{D,Z} \cong C^{2r\times(s-2r)}\cdot\{Sp(r;C) \times K\}$,   _where_

(12.30b) $K = \{(\chi;\sigma,\tau) \in L_{s-2r-a-b,a+b}(C):\ \tau \in O(a+b;C)\}.$

$\{\underline{In\ particular}\ K \cong P_{s-2r-a-b;a,b}(R)_C/\text{Skew }C^{(s-2r-a-b)\times(s-2r-a-b)}.\}$

Proof. The first assertion on the stabilizer $M_{D,Z}$ follows from
(12.13c), Lemma 12.22 and the fact that $GL(s;C)$ centralizes $O(t,u;C)$ in
$P_{s;t,u}(R)_C.$

For the isomorphism class of $M_{D,Z}$, we pass to a $GL(s;C)$-conjugate and
assume $D = \begin{bmatrix} D' & 0 \\ 0 & 0 \end{bmatrix}$ with $D' = \begin{bmatrix} 0 & I \\ -I & 0 \end{bmatrix} \in \text{Skew }C^{2r\times2r}$ and $D\cdot Z = 0$. Then
$\beta = \begin{bmatrix} a & b \\ c & d \end{bmatrix} \in GL(s;C)$ satisfies ${}^t\beta D\beta = D$ if, and only if, both ${}^ta\cdot D'\cdot a = D'$
and $b = 0$. Assuming those conditions, noting that $DZ = 0$ says $Z = \begin{bmatrix} 0 \\ Z' \end{bmatrix}$
with $Z' \in C^{(s-2r)\times(t,u)}$, and also noting
$D\cdot C^{s\times(t,u)} = \left\{ \begin{bmatrix} Z'' \\ 0 \end{bmatrix}:\ Z'' \in C^{2r\times(t,u)} \right\}$,   we calculate

$${}^t\beta\cdot Z{-}Zg = \begin{bmatrix} {}^tc\cdot Z' \\ {}^td\cdot Z'{-}Z'g \end{bmatrix};\quad so\quad {}^t\beta\cdot Z{-}Zg \in D\cdot C^{s\times(t,u)} \Leftrightarrow {}^td\cdot Z' = Z'g.$$

As in Lemma 10.1 that gives

$$M_{D,Z} \cong \left\{ \begin{bmatrix} a & 0 \\ c & d \end{bmatrix}:\ {}^ta\cdot D'\cdot a = D'\ \text{and}\ {}^td\cdot Z' \in Z'\cdot O(t,u;C) \right\}.$$

Following this by $x \mapsto {}^tx^{-1}$, now $M_{D,Z} \cong C^{2r\times(s-2r)}\cdot\{H \times K\}$ where

$$H = \{a \in GL(2r;C):\ a\cdot D'\cdot{}^ta = D'\} = Sp(r;C)$$

and

$$K = \{d \in GL(s-2r;C): d \cdot \mathcal{B}(Z',Z') \cdot {}^t d = \mathcal{B}(Z',Z')\}$$

$$\cong \left\{ \begin{bmatrix} \sigma & \chi\tau \\ 0 & \tau \end{bmatrix} \in GL(s-2r;C): \tau \in O(a+b;C) \right\}$$

$$\cong \{(\chi;\sigma,\tau) \in L_{s-2r-a-b,a+b}(C): \tau \in O(a+b;C)\}$$

where $\mathcal{B}(Z',Z')$ has rank $a+b$.

q.e.d.

The argument of Lemma 6.6 gives us

12.31. Lemma. *If* $[\pi_{D,Z,\gamma}] \in G_{s;t,u}(R)_C^\wedge$ *, then* $\pi_{D,Z,\gamma}$ *extends to a unitary representation* $\tilde{\pi}_{D,Z,\gamma}$ *of* $G_{s;t,u}(R)_C \cdot M_{D,Z}$ *on the same Hilbert space.*

Now Mackey's little-group method says that $P_{s;t,u}(R)_C^\wedge$ consists of the

$$(12.32a) \quad [\pi_{D,Z,\gamma,\mu}] = [Ind_{G_{s;t,u}(R)_C \cdot M_{D,Z} \uparrow P_{s;t,u}(R)_C}(\tilde{\pi}_{D,Z,\gamma} \otimes \mu)]$$

where, after the normalization of Lemma 12.23,

$$(12.32b) \quad D \in Skew\ C^{s \times s},\ Z \in C^{s \times (t,u)} \text{ with } {}^t\beta D\beta \cdot {}^t\beta Z = 0 \text{ for } \beta \in M_{D,Z},$$

$$(12.32c) \quad [\gamma] \in \hat{L}_{D,Z}, \text{ and}$$

$$(12.32d) \quad [\mu] \in \hat{M}_{D,Z} \text{ is extended by } \mu(D^t,Z^t,g,\beta) = \mu(\beta).$$

$[\pi_{D,Z,\gamma,\mu}] = [\pi_{D^t,Z^t,\gamma^t,\mu^t}]$ precisely when $O(t,u;C) \times GL(s;C)$ has an element (g,β) such that

(12.33a) $D' = {}^t\beta\cdot D\beta$ and $Z' - {}^t\beta\cdot Zg \in D\cdot C^{s\times(t,u)}$,

(12.33b) $g' \to \gamma(g^{-1}g'g)$ is equivalent to γ', and

(12.33c) $\beta' \to \mu(\beta^{-1}\beta'\beta)$ is equivalent to μ'.

In view of Lemma 12.29 we define

(12.34a) $J_{2r;c,a+b}(C) = C^{2r\times(a+b+c)}\cdot\{Sp(r;C) \times L^0_{c,a+b}(C)\}$

where

(12.34b) $L^0_{c,a+b}(C) = \{(\chi;\sigma,\tau) \in L_{c,a+b}(C): \tau \in O(a+b;C)\}$.

Schematically, $J_{2r;c,a+b}(C)$ consists of the matrices

$$\left[\begin{array}{c|cc} Sp(r;C) & \multicolumn{2}{c}{C^{2r\times(a+b+c)}} \\ \hline & GL(c;C) & C^{c\times(a+b)} \\ 0 & 0 & O(a+b;C) \end{array}\right] \ .$$

Now we come to the principal result for maximal parabolic subgroups of complex orthogonal groups.

12.35. Theorem. $P_{s;t,u}(R)\hat{\ }_C$ _is the disjoint union of non-empty subsets_ $S(a+b,c,d)$ _where_ a,b,c _and_ d _are non-negative integers such that_

(*) $a + b + c + d \leqslant s,\ a + c \leqslant t,\ b + c \leqslant u,$ _and_ d _is even._

Given $(a+b,c,d)$ _satisfying_ (*), _define_ $D = \begin{bmatrix} D' & 0 \\ 0 & 0 \end{bmatrix} \in$ Skew $C^{s\times s}$ _where_

$$D' = \begin{bmatrix} 0 & I \\ -I & 0 \end{bmatrix} \in \text{Skew } C^{d \times d}. \quad \underline{Choose} \quad Z \in C^{s \times (t,u)} \quad \underline{such} \ \underline{that}$$

$$DZ = 0 \ , \ \mathcal{B}(Z,Z) \ \underline{has} \ \underline{rank} \ a{+}b \ \underline{and} \ Z \ \underline{has} \ \underline{rank} \ a{+}b{+}c.$$

$\underline{Then}$ $S(a{+}b,c,d)$ $\underline{is \ parameterized \ by}$

$$G_{c;t-c-a,u-c-b}(R)_C^{\wedge} \times J_{d;s-d-a-b,a+b}(C)^{\wedge}$$

$\underline{under}$

$$([\gamma],[\mu]) \leftrightarrow [\pi_{D,Z,\gamma,\mu}] \quad \underline{given \ by} \quad (12.32)$$

$\underline{where}$

$$L_{D,Z} \quad \underline{is \ identified \ with \ its \ isomorph} \quad G_{c;t-c-a,u-c-b}(R)_C$$

$\underline{and}$

$$M_{D,Z} \quad \underline{is \ identified \ with \ its \ isomorph} \quad J_{d;s-d-a-b,a+b}(C).$$

At this point, verification of the parameterization is routine. The $G_{c;t-c-a,u-c-b}(R)^{\wedge}$ were described in Theorem 12.28. We now go on to the determination of the $J_{d;s-d-a-b,a+b}(C)^{\wedge}$.

We view $J_{2r;c,a+b}(C)$ as the semidirect product

$$(12.36a) \qquad J_{2r;c,a+b}(C) = I_{2r;c+a+b}(C) \cdot L_{c,a+b}^0(C)$$

where $L_{c,a+b}^0(C) = \{(\chi;\sigma,\tau) \in L_{c,a+b}(C): \tau \in O(a{+}b;C)\}$, as in (12.34b), and where

(12.36b) $I_{2r;c+a+b}(C) = C^{2r\times(c+a+b)} \cdot Sp(r;C)$.

Looking back to (8.5) we see

(12.37a) $I_{2r;c+a+b}(C) \cong G_{c+a+b;2r}(C)/Sym\ C^{(c+a+b)\times(c+a+b)}$.

$Sym\ C^{(c+a+b)\times(c+a+b)}$ is central in the unipotent radical $N_{c+a+b;2r}(C)$ of $G_{c+a+b;2r}(C)$. In the notation (9.29) and (9.30), that tells us

(12.37b) $I_{2r;c+a+b}(C)\hat{} = \{[\pi_{D,Z,\gamma}] \in G_{c+a+b;2r}(C)\hat{}: D = 0\}$.

Now Theorem 9.33 specializes to

12.38. Proposition. $I_{2r;c+a+b}(C)\hat{}$ *is the disjoint union of non-empty subsets* $S_I(k,\ell)$, *where* k *and* ℓ *are non-negative integers such that*

(*) $k + 2\ell \leq c + a + b$ *and* $k + \ell \leq r$,

as follows. Let $\mathcal{A}_\ell$ *denote the space of all complex* $(c+a+b) \times (c+a+b)$ *antisymmetric matrices* A *of rank* 2ℓ. *Then* $S_I(k,\ell)$ *is parameterized by* $\mathcal{A}_\ell \times G_{k;2(r-k-\ell)}(C)\hat{}$ *under*

          $(A,[\gamma]) \leftrightarrow [\pi_{0,Z,\gamma}]$ *in the notation* (9.29), (9.30), (12.37)

where $Z \in C^{(c+a+b)\times 2r}$ *with* $\mathcal{B}(Z,Z) = A$ *and* rank $Z = k + 2\ell$, $\mathcal{B}$ *given by* (9.32).

The $L^0_{c,a+b}(C)$-stabilizer of $[\pi_{0,Z,\gamma}] \in I_{2r;c+a+b}(C)\hat{}$ is obtained by specializing Lemma 10.1 to the case $D = 0$. If $Z \in C^{(c+a+b)\times 2r}$ with rank $Z = k+2\ell$ and rank $\mathcal{B}(Z,Z) = 2\ell$, the stabilizer is

(12.39a) $M_Z = \{\beta \in L^0_{c,a+b}(\mathbb{C}) \subset GL(c+a+b;\mathbb{C}): {}^t\beta \cdot Z \in Z \cdot Sp(r;\mathbb{C})\}$.

Restrict the extension of Lemma 10.3 to see that

(12.39b)   every $[\pi_{0,Z,\gamma}] \in I_{2r;c+a+b}(\mathbb{C})^\wedge$ extends to $I_{2r;c+a+b}(\mathbb{C}) \cdot M_Z$.

Now (little-group method)

12.40. Theorem. $J_{2r;c,a+b}(\mathbb{C})^\wedge$ _is the disjoint union of non-empty_
subsets $S_J(k;\{A\})$ _where_ k _and_ ℓ _are non-negative integers such that_

$$k + 2\ell \leqslant c + a + b \quad \underline{and} \quad k + \ell \leqslant r$$

and where $\{A\}$ _is an_ $L^0_{c,a+b}(\mathbb{C})$ _-equivalence class of_ $(c+a+b) \times (c+a+b)$
antisymmetric complex matrices A _of rank_ 2ℓ. $S_J(k;\{A\})$ _is parameterized by_
$G_{k;2(r-k-\ell)}(\mathbb{C})^\wedge \times \hat{M}_Z$ _under_

$$([\gamma],[\mu]) \leftrightarrow [\mathrm{Ind}_{I_{2r;c+a+b}(\mathbb{C}) \cdot M_Z \uparrow J_{2r;c,a+b}(\mathbb{C})}(\tilde{\pi}_{0,Z,\gamma} \otimes \mu)]$$

where $Z \in \mathbb{C}^{(c+a+b) \times 2r}$ _with_ $\mathcal{B}(Z,Z) = A$ _and_ rank $Z = k + 2\ell$ _with_ $\mathcal{B}$
given by (9.32).

   This reduces the specification of $J_{2r;c,a+b}(\mathbb{C})^\wedge$ to that of the $\hat{M}_Z$.
As we are about to see, the M_Z are groups of similar type, constructed
from smaller matrices. By recursion on the degree of the matrices we thus
may assume the $\hat{M}_Z$ known.

   To complete our discussion of $J_{2r;c,a+b}(\mathbb{C})$, we note the $\mathbb{C}^{c+a+b}$
carries

(12.41a)   an antisymmetric bilinear form α of rank 2ℓ
and

(12.41b)   a symmetric bilinear form φ of rank a+b

such that

(12.41c) $M_Z = \{g \in GL(c+a+b;C): g \text{ preserves both } \alpha \text{ and } \varphi\}$.

For $L^0_{c,a+b}(C) = \{g \in GL(c+a+b;C): g \text{ preserves } \varphi\}$, and the condition
(12.39a) that $^tg{\cdot}Z \in Z{\cdot}Sp(r;C)$ just says that g preserves the antisymmetric form α with matrix $\mathcal{B}(Z,Z)$, $\mathcal{B}$ given by (9.32). Decompose
$C^{c+a+b} = U \oplus V \oplus W$ where

$U \oplus V$ is the null space of φ, so $\varphi|_{W\times W}$ is nondegenerate,

U is the null space of $\alpha|_{(U+V)\times(U+V)}$, so $\alpha|_{V\times V}$ is nondegenerate,

W is orthogonal to V relative to α.

Every $g \in M_Z$ preserves $U \oplus V$ and U, and sends W to $U \oplus W$. Now
$g \in M_Z$ is given schematically by

$$\begin{bmatrix} g_{11} & g_{12} & g_{13} \\ 0 & g_{22} & 0 \\ 0 & 0 & g_{33} \end{bmatrix} \in \begin{bmatrix} GL(U) & U \oplus V^* & U \oplus W^* \\ 0 & Sp(\alpha|_{V\times V}) & 0 \\ 0 & 0 & 0(\varphi|_{W\times W}) \end{bmatrix}$$

where g_{13} and g_{33} are connected by the condition that $\alpha(gw,gw') = \alpha(w,w')$
for all $w,w' \in W$. Set $n_1 = \dim U$, $n_2 = \dim V$, and $n_3 = \dim W$. Then
$M_Z \cong I'_{n_1,n_2;n_3}(C){\cdot}L^{Sp}_{n_1,n_2}(C)$ where $L^{Sp}_{n_1,n_2}(C) = \{(\chi;\sigma,\tau) \in L_{n_1,n_2}(C):$
$\tau \in Sp(\frac{1}{2}n_2;C)\}$ as in (10.7c) and where $I'_{n_1,n_2;n_3}(C)$ is the subgroup
$(g|_{U+V} = \text{identity})$ in M_Z. We view $I'_{n_1,n_2;n_3}(C)$

$$\begin{bmatrix} I & 0 & g_{13} \\ 0 & I & 0 \\ 0 & 0 & g_{33} \end{bmatrix} \rightarrow \begin{bmatrix} I & g_{13} \\ 0 & g_{33} \end{bmatrix} \rightarrow (g_{13}\, g_{33}^{-1}\,,\, g_{33})$$

in $C^{n_1 \times n_3} \cdot O(n_3; C)$. Given specific α and φ, now we know enough to determine $\hat{M}_Z$.

This completes our determination of the unitary duals of the maximal parabolic subgroups of the complex orthogonal groups.

§13.   Structure of Parabolic Subgroups of $SO^*(2m)$.

In this section we come to the last series of real classical groups.
Those are the groups $SO^*(2m)$, real forms of the complex special (determinant 1) orthogonal groups $SO(2m;C)$ in which the maximal compact subgroups
are isomorphic to $U(m)$. The groups $SO^*(2m)$ occur as automorphism groups
for one of the four classical series of bounded symmetric domains (see [27,
§12]).

In the sequel we use both the symmetric bilinear form on C^{2m} given by

$$(13.1a) \qquad \sigma(x,y) = \sum_{1 \leqslant \ell \leqslant m} (x^\ell y^{m+\ell} + x^{m+\ell} y^\ell)$$

and the hermitian form of signature (m,m) given by

$$(13.1b) \qquad h(x,y) = \sum_{1 \leqslant \ell \leqslant m} x^\ell \bar{y}^\ell - \sum_{1 \leqslant \ell \leqslant m} x^{m+\ell} \bar{y}^{m+\ell} .$$

The orthogonal group of σ is $O(2m;C)$, the unitary group of h is $U(m,m)$,
and we are interested in the group

$$(13.1c) \quad SO^*(2m) = O(2m;C) \cap U(m,m)$$

$$= \{g \in GL(2m;C): \ g \text{ preserves both } \sigma \text{ and } h\} .$$

Let us check that this really is the desired real form of $SO(2m;C)$:

13.2. Lemma. *The group* $SO^*(2m)$ *defined in* (13.1) *is the real form
of* $SO(2m;C)$ *defined by the involution* $g \to \begin{bmatrix} I & 0 \\ 0 & -I \end{bmatrix} \cdot (g^{-1})^* \cdot \begin{bmatrix} I & 0 \\ 0 & -I \end{bmatrix}$. $SO^*(2m)$
has Cartan involution $\theta(g) = (g^{-1})^*$, *and the corresponding maximal compact
subgroup is*

$$\left\{ g \in SO^*(2m): \theta(g) = g \right\} = \left\{ \begin{bmatrix} k & 0 \\ 0 & \bar{k} \end{bmatrix} : k \in U(m) \right\} \cong U(m).$$

Finally, $SO^*(2m)$ *has Lie algebra*

$$\mathfrak{so}^*(2m) = \left\{ \begin{bmatrix} A & B \\ -\bar{B} & \bar{A} \end{bmatrix} : A + A^* = 0 = B + {}^tB \right\}.$$

$\underline{Proof}$: Let $\tau: GL(2m;\mathbb{C}) \to GL(2m;\mathbb{C})$ by $\tau(g) = \begin{bmatrix} I & 0 \\ 0 & -I \end{bmatrix} \cdot (g^{-1})^* \cdot \begin{bmatrix} I & 0 \\ 0 & -I \end{bmatrix}$.
Then τ is a conjugate linear involution of $GL(2m;\mathbb{C})$ with fixed point
set $\left\{ g \in GL(2m;\mathbb{C}): g^* \cdot \begin{bmatrix} I & 0 \\ 0 & -I \end{bmatrix} \cdot g = \begin{bmatrix} I & 0 \\ 0 & -I \end{bmatrix} \right\} = U(m,m)$. $O(2m;\mathbb{C})$ is τ-stable
because $\tau \begin{bmatrix} I & 0 \\ 0 & -I \end{bmatrix} = \begin{bmatrix} I & 0 \\ 0 & -I \end{bmatrix}$. Now $SO^*(2m)$ is the real form of $O(2m;\mathbb{C})$
defined by the restriction of τ.

$\theta(g) = (g^{-1})^*$ is a Cartan involution of $GL(2m;\mathbb{C})$. $SO^*(2m)$ is
θ-stable, so θ restricts to a Cartan involution of $SO^*(2m)$. Now $SO^*(2m)$
has maximal compact subgroup $\{g \in SO^*(2m): \theta(g) = g\} = SO^*(2m) \cap U(2m)$,
which is

$$O(2m;\mathbb{C}) \cap \{U(m,m) \cap U(2m)\}$$

$$= O(2m;\mathbb{C}) \cap \left\{ \begin{bmatrix} k' & 0 \\ 0 & k'' \end{bmatrix} : k',k'' \in U(m) \right\}$$

$$= \left\{ g = \begin{bmatrix} k' & 0 \\ 0 & k'' \end{bmatrix} : k^{(i)} \in U(m) \text{ and } {}^tg \cdot \begin{bmatrix} 0 & I \\ I & 0 \end{bmatrix} \cdot g = \begin{bmatrix} 0 & I \\ I & 0 \end{bmatrix} \right\}$$

$$= \left\{ \begin{bmatrix} k' & 0 \\ 0 & k'' \end{bmatrix} : k'' = ({}^tk')^{-1} \in U(m) \right\} = \left\{ \begin{bmatrix} k & 0 \\ 0 & \bar{k} \end{bmatrix} : k \in U(m) \right\}.$$

This is a connected group, isomorphic to $U(m)$, so $SO^*(2m)$ is connected.
Now $SO^*(2m)$ is contained in the identity component $SO(2m;\mathbb{C})$ of $O(2m;\mathbb{C})$.
This completes the proof of the group level statements. The Lie algebra
statement is routine.

$q.e.d.$

The Lie group $SO^*(2m)$ has real rank $[m/2]$, and C^{2m} has $[m/2]$ $SO^*(2m)$-equivalence classes of nonzero even-dimensional σ-isotropic h-isotropic subspaces. In the basis $\{e_1,\ldots,e_{2m}\}$ used for (13.1), these equivalence classes of doubly isotropic subspaces are represented by the

(13.3a) E_{2s}: span of $\{u_1,\ldots,u_s;v_1,\ldots,v_s\}$, $1 \leqslant s \leqslant [m/2]$,

where

(13.3b) $u_t = -e_{2t} + e_{n+2t-1}$ and $v_t = e_{2t-1} + e_{n+2t}$.

The maximal parabolic subgroups of $SO^*(2m)$ are the subgroups specified by the E_{2s} and their conjugates as follows.

      13.4. Lemma. *The maximal parabolic subgroups of* $SO^*(2m)$ *are the*

$$P_E = \{g \in SO^*(2m): gE = E\}$$

where

      $E \subset C^{2m}$ *has dimension* $2s > 0$ *with* $\sigma(E,E) = 0 = h(E,E)$.

In particular, $SO^*(2m)$ *has exactly* $[m/2]$ *conjugacy classes of maximal parabolic subgroups, represented by the* $P_{E_{2s}}$ *for* $s = 1,2,\ldots,[m/2]$.

      Proof. This can be proved directly by some linear algebra, but here is a noncomputational proof. Let X_0 be the bounded symmetric domain $SO^*(2m)/U(m)$. As symmetric space it has rank $[m/2]$, so [27, §5] there are exactly $[m/2]$ equivalence classes of boundary components. Furthermore [27, §6] the maximal parabolic subgroups of $SO^*(2m)$ are just the $SO^*(2m)$-normalizers of boundary components of X_0. Finally [27, §12] the just-described groups P_E are normalizers of boundary components.

q.e.d.

If s and t are non-negative integers, we define real-bilinear maps

(13.5a) $\mathcal{L}, \mathcal{M}: C^{2s \times t} \times C^{2s \times t} \to C^{2s \times 2s}$

as follows. Let $p = \begin{bmatrix} 0 & I \\ I & 0 \end{bmatrix}$ and $q = \begin{bmatrix} -I & 0 \\ 0 & I \end{bmatrix}$ in $s \times s$ blocks. Then

(13.5b) $\mathcal{M}(Z, Z') = Z \cdot (qZ')^* + p \cdot {}^t\{Z' \cdot (qZ)^*\} \cdot p$ and

$\mathcal{L}(Z, Z') = \frac{1}{2}\{\mathcal{M}(Z, Z') - \mathcal{M}(Z', Z)\}.$

One checks that ${}^t\mathcal{M}(Z, Z') \cdot p + p \cdot \mathcal{M}(Z, Z')$ and $\mathcal{M}(Z, Z')^* \cdot q + q \cdot \mathcal{M}(Z, Z')$ are symmetric in Z and Z'. As $\mathcal{L}$ is alternating, now

(13.5c) $\mathcal{L}$ takes values in the Lie algebra $\mathfrak{so}^*(2s)$,

and $\mathcal{L}$ defines a Lie group and its Lie algebra by the formulae

(13.6a) $N^*_{2s;t} = SO^*(2s) + C^{2s \times t}$ with $(D, Z)(D', Z') = (D + D' + \mathcal{L}(Z, Z'), Z + Z')$

and

(13.6b) $n^*_{2s;t} = \mathfrak{so}^*(2s) + C^{2s \times t}$ with $[(D, Z), (D', Z')] = (2\mathcal{L}(Z, Z'), 0).$

$SO^*(2t)$ acts on $N^*_{2s;t}$ by the formula

(13.7a) $g(D, Z) = (D, Z_g)$ where $(Z_g, pq\overline{Z_g}) = (Z, pq\overline{Z}) \cdot g^{-1}.$

If we observe that, where $\tilde{q}$ has the form of q in $t \times t$ blocks,

(13.7b) $\mathcal{M}(Z,Z') = Z \cdot Z'^* \cdot g - (pq\bar{Z}) \cdot (pq\bar{Z}')^* \cdot q = (Z,pq\bar{Z}) \cdot \tilde{q} \cdot (Z',pq\bar{Z}')^* \cdot q$

then we see

(13.7c) $\mathcal{M}(Z_g, Z'_g) = \mathcal{M}(Z,Z')$ for $g \in U(t,t)$, thus for $g \in SO^*(2t)$.

Now $SO^*(2t)$ acts by automorphisms on $N^*_{2s;t}$ and we have semidirect product groups

(13.7d) $G^*_{2s;t} = N^*_{2s;t} \cdot SO^*(2t)$

View the quaternion general linear group as

(13.8a) $GL(s;Q) = \left\{ \gamma \in GL(2s;C) : \begin{bmatrix} 0 & I \\ -I & 0 \end{bmatrix} \cdot \gamma \cdot \begin{bmatrix} 0 & I \\ -I & 0 \end{bmatrix}^{-1} = \bar{\gamma} \right\}$,

that is C^{2s} is viewed as Q^s where $x \to \begin{bmatrix} 0 & I \\ -I & 0 \end{bmatrix} \bar{x}$ is scalar multiplication by j. Then $GL(s;Q)$ acts on $N^*_{2s;t}$ by

(13.8b) $\gamma(D,Z) = (\gamma D \gamma'^{-1}, \gamma Z)$ where $\gamma' = p \cdot {}^t\gamma^{-1} \cdot p$.

One checks that

(13.8c) $\mathcal{M}(\gamma Z, \gamma Z^t) = \gamma \cdot \mathcal{M}(Z, Z^t) \cdot \gamma'^{-1}$ for $\gamma \in GL(s;Q)$.

Thus $GL(s;Q)$ acts by automorphisms on $N^*_{2s;t}$. This action commutes with the action of $SO^*(2t)$, so we have semidirect product groups

(13.8d) $P^*_{2s;t} = G^*_{2s;t} \cdot GL(s;Q) = N^*_{2s;t} \cdot \{SO^*(2t) \times GL(s;Q)\}$,

with group law

$$(D_0,Z_0;g_0,\gamma_0)(D,Z;g,\gamma) = (D_0+\gamma_0 D\gamma_0'^{-1}+\mathcal{L}(Z_0,\gamma_0 \cdot Z_{g_0}), Z_0+\gamma_0 \cdot Z_{g_0}; g_0 g, \gamma_0 \gamma).$$

We finally come to the structure theorem for the maximal parabolic subgroups of our groups $SO^*(2m)$:

   13.9. Theorem. *Let* E *be a totally* σ-*isotropic and* h-*isotropic subspace of even dimension* $2s > 0$ *in* C^{2m}, *and let* $P_E \subset SO^*(2m)$ *be the corresponding maximal parabolic subgroup. Then there is a Lie group/algebraic group isomorphism*

(13.10a) $\varphi : P^*_{2s;m-2s} \to P_E = \{g \in SO^*(2m): gE = E\}$

that carries $N^*_{2s;m-2s}$ *to the unipotent radical, carries* $SO^*(2m-4s) \times GL(s;Q)$ *to a reductive complement, and restricts to an isomorphism*

(13.10b) $\varphi : G^*_{2s;m-2s} \to L_E = \{g \in SO^*(2m): g|_E = identity\}$.

The Langlands decomposition P_E = MAN *is reflected by*

(13.10c)   M = $\varphi \cdot \{SO^*(2m-4s) \times GL'(s;Q)\}$,   N = $\varphi \cdot N^*_{2s;m-2s}$ *and* A = $\varphi \cdot R^+$.

Some immediate consequences:

   13.11. Corollary. *A maximal parabolic subgroup* $P_E \subset SO^*(2m)$ *has abelian unipotent radical if and only if* dim E = m.

   13.12. Corollary. *A maximal parabolic subgroup* $P_E \subset SO^*(2m)$ *is cuspidal if and only if* dim E = 2.

We turn to the proof of Theorem 13.9. Let $\{e_1,\ldots,e_{2m}\}$ be the basis of C^{2m} in which σ and h are given by (13.1). We may suppose that $E = E_{2s}$ as in (13.3), that is E is spanned by the

$$u_i = -e_{2i} + e_{m+2i-1} \quad \text{and} \quad v_i = e_{2i-1} + e_{m+2i}$$

for $1 \leqslant i \leqslant s$. Set

$$(13.13a) \qquad u_i' = \tfrac{1}{2}(e_{2i} + e_{m+2i-1}) \quad \text{and} \quad v_i' = \tfrac{1}{2}(e_{2i-1} - e_{m+2i})$$

and observe that

$$(13.13b) \qquad \sigma(u_i,v_j') = \delta_{ij} = \sigma(v_i,u_j') \quad \text{and} \quad \sigma(u_i,u_j') = 0 = \sigma(v_i,v_j')$$

and

$$(13.13c) \qquad -h(u_i,u_j') = \delta_{ij} = h(v_i,v_j') \quad \text{and} \quad h(u_i,v_j') = 0 = h(v_i,u_j') \ .$$

Now define

$$(13.14a) \qquad E' = \text{span}\ \{u_1',\ldots,u_s';\ v_1',\ldots,v_s'\}\ ,$$

$$(13.14b) \qquad V = E + E' = \text{span}\ \{e_1,\ldots,e_{2s};\ e_{m+1},\ldots,e_{m+2s}\},$$

$$(13.14b) \qquad W = V^\perp = \text{span}\ \{e_{2s+1},\ldots,e_m;\ e_{m+2s+1},\ldots,e_{2m}\}.$$

Then $C^{2m} = V \oplus W,$ orthogonal under both σ and h. Denote

$$(13.15a) \qquad \mathfrak{p} = \{\xi \in \mathfrak{so}^*(2m)\colon \xi E \subseteq E\}\ , \quad \text{Lie algebra of } P_E \text{ and}$$

(13.15b) $\mathfrak{l} = \{\xi \in \mathfrak{so}^*(2m): \xi E = 0\}$, Lie algebra of L_E.

As before, the first step in finding the structure of P_E and L_E is

   13.16. Lemma. $\mathfrak{p}$ *is direct sum of its linear subspaces*

$$\mathfrak{p}_r^W = \{\xi \in \mathfrak{so}^*(2m): \xi V = 0 \underline{\quad and \quad} \xi W \subset W\},$$

$$\mathfrak{p}_r^V = \{\xi \in \mathfrak{so}^*(2m): \xi E \subset E, \xi E' \subset E' \underline{\quad and \quad} \xi W = 0\},$$

$$\mathfrak{p}_n^1 = \{\xi \in \mathfrak{so}^*(2m): \xi E' \subset W, \xi W \subset E \underline{\quad and \quad} \xi E = 0\}, \underline{\quad and \quad}$$

$$\mathfrak{p}_n^2 = \{\xi \in \mathfrak{so}^*(2m): \xi E' \subset E \underline{\quad and \quad} \xi(E + W) = 0\};$$

and $\mathfrak{l} = \mathfrak{p}_r^W + \mathfrak{p}_n^1 + \mathfrak{p}_n^2.$

   Proof. Lemma 3.4 applies to the unitary group $U(m,m)$ of h and gives the corresponding decompositions

$$'\mathfrak{p} = '\mathfrak{p}_r^W + '\mathfrak{p}_r^V + '\mathfrak{p}_n^1 + '\mathfrak{p}_n^2 \text{ and } '\mathfrak{l} = '\mathfrak{p}_r^W + '\mathfrak{p}_n^1 + '\mathfrak{p}_n^2$$

of the Lie algebras of

$$'P = \{g \in U(m,m): gE = E\} \text{ and } 'L = \{g \in U(m,m): g|_E = \text{identity}\}.$$

Complexifying, it also applies to the orthogonal group $O(2m;\mathbb{C})$ of σ and gives the corresponding decompositions

$$''\mathfrak{p} = ''\mathfrak{p}_r^W + ''\mathfrak{p}_r^V + ''\mathfrak{p}_n^1 + ''\mathfrak{p}_n^2 \text{ and } ''\mathfrak{l} = ''\mathfrak{p}_r^V + ''\mathfrak{p}_n^1 + ''\mathfrak{p}_n^2$$

of the Lie algebras of

152 Joseph A. Wolf

$''P = \{g \in 0(2m;C): gE = E\}$ and $''L = \{g \in 0(2m;C): g|_E = \text{identity}\}.$

Now

$$\mathfrak{p} = '\mathfrak{p} \cap ''\mathfrak{p} \qquad \text{and} \qquad \mathfrak{l} = '\mathfrak{l} \cap ''\mathfrak{l}$$

decompose as sums of intersections of the corresponding pieces, and this gives our assertion.

$$\underline{q.e.d.}$$

Now we identify the group corresponding to $\mathfrak{p}_r^V$.

13.17. **Lemma.** *Let* $H = \{g: V \to V$ *linear:* $gE = E$, $gE' = E'$, *and* g *preserves both* $\sigma|_{V \times V}$ *and* $h|_{V \times V}\}$. *Then* H *consists of all linear transformations* $g: V \to V$ *such that*

$$g \text{ has matrix } \begin{bmatrix} \gamma & 0 \\ 0 & \gamma' \end{bmatrix} \text{ in the basis}$$

$$\{u_1,\ldots,u_s,v_1,\ldots,v_s;u_1^t,\ldots,u_s^t,v_1^t,\ldots,v_s^t\}$$

where

$$\gamma' = \begin{bmatrix} 0 & I \\ I & 0 \end{bmatrix} \cdot {}^t\gamma^{-1} \cdot \begin{bmatrix} 0 & I \\ I & 0 \end{bmatrix} \quad \underline{and} \quad \begin{bmatrix} 0 & I \\ -I & 0 \end{bmatrix} \cdot \gamma \cdot \begin{bmatrix} 0 & I \\ -I & 0 \end{bmatrix}^{-1} = \bar{\gamma}.$$

View C^{2s} *as a quaternionic vector space* Q^s *where scalar multiplication by* j *is* $x \to \begin{bmatrix} 0 & I \\ -I & 0 \end{bmatrix} \cdot \bar{x}$. *Then the condition* $\begin{bmatrix} 0 & I \\ -I & 0 \end{bmatrix} \gamma \begin{bmatrix} 0 & I \\ -I & 0 \end{bmatrix}^{-1} = \bar{\gamma}$ *just says* $\gamma \in GL(s;Q)$. *In particular,* $H \cong GL(s;Q)$.

Proof. Let $g: V \to V$ linear such that $gE = E'$ and $gE' = E'$. In the basis $\{u_1,\ldots,v_s,u_1',\ldots,v_s'\}$, g has matrix of the form $\begin{bmatrix} \gamma & 0 \\ 0 & \gamma' \end{bmatrix}$. Denote $p = \begin{bmatrix} 0 & I \\ I & 0 \end{bmatrix}$ and $q = \begin{bmatrix} -I & 0 \\ 0 & I \end{bmatrix}$ in $s \times s$ blocks. Then (13.13) tells us that g preserves $\sigma|_{V \times V} \Leftrightarrow \begin{bmatrix} {}^t\gamma & 0 \\ 0 & {}^t\gamma' \end{bmatrix} \cdot \begin{bmatrix} 0 & p \\ p & 0 \end{bmatrix} \cdot \begin{bmatrix} \gamma & 0 \\ 0 & \gamma' \end{bmatrix} = \begin{bmatrix} 0 & p \\ p & 0 \end{bmatrix}$

and

g preserves $h|_{V \times V} \Leftrightarrow \begin{bmatrix} \gamma^* & 0 \\ 0 & \gamma'^* \end{bmatrix} \cdot \begin{bmatrix} 0 & q \\ q & 0 \end{bmatrix} \cdot \begin{bmatrix} \gamma & 0 \\ 0 & \gamma' \end{bmatrix} = \begin{bmatrix} 0 & q \\ q & 0 \end{bmatrix}$.

In other words $g \in H \Leftrightarrow {}^t\gamma \cdot p\gamma' = p$ and $\gamma^* \cdot q\gamma' = q$, that is
$g \in H \Leftrightarrow \gamma' = p \cdot {}^t\gamma^{-1} \cdot p$ and $\bar{\gamma} = q \cdot {}^t\gamma'^{-1} \cdot q = (qp)\gamma(qp)^{-1}$, as asserted. The
other assertions follow. $\hspace{4cm}$ *q.e.d.*

This gives us the group corresponding to $\mathfrak{p}_r = \mathfrak{p}_r^W \oplus \mathfrak{p}_r^V$:

13.18. Lemma. *Define* $\varphi_r \colon SO^*(2m-4s) \times GL(s;\mathbb{Q}) \to SO^*(2m)$ *by*

(i) $\varphi_r|_{SO^*(2m-4s)}$ *is the isomorphism of* $SO^*(2m-4s)$ *onto*
the corresponding group $\{g \in SO^*(2m): g|_V = 1$ *and* $gW = W\}$ *of* W *that is*
induced by the orthogonal/unitary isomorphism

$$x \to \sum_{\ell=1}^{m-2s} (e_{2s+\ell}x^\ell + e_{m+2s+\ell}x^{m+2s+\ell})$$

(ii) $\varphi_r|_{GL(s;\mathbb{Q})}$ *is the isomorphism of* $GL(s;\mathbb{Q})$ *onto*
$\{g \in SO^*(2m): gE = E, gE' = E'$ *and* $g|_W = 1\}$ *specified by Lemma* 13.17.

Then φ_r *is an isomorphism of* $SO^*(2m-4s) \times GL(s;\mathbb{Q})$
$\{g \in P_E: g\mathfrak{p}_r g^{-1} = \mathfrak{p}_r\}$.

Proof. By construction, φ_r induces an isomorphism of
$SO^*(2m-4s) \times GL(s;\mathbb{Q})$ onto its image, and image(φ_r) is a closed subgroup
of P_E with Lie algebra $\mathfrak{p}_r$. So we need only verify that image(φ_r)
contains the P_E-normalizer of $\mathfrak{p}_r$.

If $g \in P_E$ normalizes $\mathfrak{p}_r$, the result of Lemma 3.13 for $U(m,m)$ shows
$gV = V$ and $gW = W$, and so $g = g_V g_W$ where $g_V(v+w) = g(v) + w$ and
$g_W(v+w) = v + g(w)$. It follows that $g \in$ image(φ_r). $\hspace{2cm}$ *q.e.d.*

Let $p = \begin{bmatrix} 0 & I \\ I & 0 \end{bmatrix}$ and $q = \begin{bmatrix} -I & 0 \\ 0 & I \end{bmatrix}$ as in the proof of Lemma 13.17. We write matrices in the basis

$$(13.19a) \quad \{u_1,\ldots,u_s,v_1,\ldots,v_s;u_1',\ldots,u_s';v_1',\ldots,v_s';e_{2s+1},\ldots,e_m,e_{m+2s+1},\ldots,e_{2m}\}$$

of C^{2m}. Let $\eta,\zeta : C^{2m} \to C^{2m}$ linear such that $\eta E' \subset W$, $\eta W \subset E$, $\eta E = 0$; $\zeta E' \subset E$, and $\zeta(E+W) = 0$. Then in the basis (13.19a) they have matrices of the form

$$\eta = \begin{bmatrix} 0 & 0 & Z & W \\ 0 & 0 & 0 & 0 \\ 0 & X & 0 & 0 \\ 0 & Y & 0 & 0 \end{bmatrix} \quad \text{and} \quad \zeta = \begin{bmatrix} 0 & -D & 0 & 0 \\ 0 & 0 & 0 & 0 \\ 0 & 0 & 0 & 0 \\ 0 & 0 & 0 & 0 \end{bmatrix}$$

Using (13.13) a direct calculation shows that

$$\zeta \in \mathfrak{so}^*(2m) \quad \text{if and only if} \quad D \in \mathfrak{so}^*(2s), \quad \text{and}$$

$$\eta \in \mathfrak{so}^*(2m) \quad \text{if and only if} \quad W = pq\bar{Z}, \quad X = -Z^*q \quad \text{and} \quad Y = -{}^tZ\cdot p.$$

In other words,

$$(13.19b) \quad \mathfrak{p}_n^1 = \{\eta_Z : Z \in C^{2s\times(m-2s)}\} \quad \text{and} \quad \mathfrak{p}_n^2 = \{\zeta_D : D \in \mathfrak{so}^*(2s)\}$$

where

$$(13.19c) \quad \eta_Z = \begin{bmatrix} 0 & 0 & Z & pq\bar{Z} \\ 0 & 0 & 0 & 0 \\ 0 & -Z^*q & 0 & 0 \\ 0 & -{}^tZp & 0 & 0 \end{bmatrix} \quad \text{and} \quad \zeta_D = \begin{bmatrix} 0 & -D & 0 & 0 \\ 0 & 0 & 0 & 0 \\ 0 & 0 & 0 & 0 \\ 0 & 0 & 0 & 0 \end{bmatrix}$$

Recall (13.5b) and compute $\eta_Z\eta_{Z'} = \zeta_{\mathcal{M}(Z,Z')}$. Notice $\eta_Z\zeta_D = \zeta_D\eta_Z = \zeta_D\zeta_{D'} = 0$. Now

(13.20a) $\mathfrak{p}_n = \mathfrak{p}_n^1 + \mathfrak{p}_n^2$ consists of nilpotent matrices,

(13.20b) $[n_Z, n_{Z'}] = \zeta_{2\mathcal{L}(Z,Z')}$ and $\mathfrak{p}_n^2$ is central in $\mathfrak{p}_n$;

and so

(13.20c) $\psi_n(D,Z) = n_Z + \zeta_D$ is an isomorphism of $\overset{*}{\mathfrak{n}}_{2s;m-2s}$ onto $\mathfrak{p}_n$.

    Recall the injective homomorphism φ_r form $SO^*(2m-4s) \times GL(s;Q)$ to
$SO^*(2m)$ given by Lemma 13.18. In the basis (13.19a) it is given by

(13.21a) $\varphi_r(g,\gamma) = \left[\begin{array}{cc|c} \gamma & 0 & \\ 0 & \gamma^{\iota} & 0 \\ \hline & 0 & g \end{array}\right]$ where $\gamma' = \begin{bmatrix} 0 & I \\ I & 0 \end{bmatrix} \cdot {}^t\gamma^{-1} \cdot \begin{bmatrix} 0 & I \\ I & 0 \end{bmatrix}$.

Now calculate

(13.21b) $\varphi_r(g,\gamma) \cdot \psi_n(D,Z) \cdot \varphi_r(g,\gamma)^{-1} = \psi_n(\gamma D \gamma^{\iota-1}, \gamma Z'_g)$

If we define

(13.21c) $\varphi_n \colon \overset{*}{N}_{2s;m-2s} \to SO^*(2m)$,   homomorphism that induces ψ_n,

then (13.21b) remains valid with ψ_n replaced by φ_n. This proves that
the map $\varphi \colon P^*_{2s;m-2s} \to SO^*(2m)$, defined by $\varphi(D,Z;g,\gamma) = \varphi_n(D,Z) \cdot \varphi_r(g,\gamma)$,
is a Lie group isomorphism onto its image. $\varphi \cdot \overset{*}{N}_{2s;m-2s}$ is the unipotent
radical of P_E by (13.20), (13.21) and Lemma 13.18, and now Lemma 13.18
says that $\varphi \cdot \{SO^*(2m-4s) \times GL(s;Q)\}$ is a rational reductive complement to
that unipotent radical. This completes the proof of Theorem 13.9.

                                                            $\underline{q.e.d.}$

§14. The Nilradical and the Intermediate Group for $SO^*(2m)$.

We work out the irreducible unitary representations of the groups $N^*_{2s;t}$ and $G^*_{2s;t}$ of §13. This follows the lines of §§4 and 5, except for some algebraic complications.

In this section we use the notation

(14.1a) $N = N^*_{2s;t}$, our simply connected nilpotent group;

(14.1b) $n = n^*_{2s;t} = so^*(2s) + C^{2s \times t}$, its Lie algebra;

(14.1c) $n^* = $ the real linear dual of n; and

(14.1d) $Ad^* = $ the co-adjoint representation of N on n^*.

We define a real pairing on n by the formula

(14.2a) $<(D,Z),(D',Z')> = $ trace Re DD' + trace Re $M(Z,Z')$.

If $0 \neq (D,Z) \in n$, then $(D^*,qZ) \in n$ and we calculate

$$<(D,Z),(D^*,qZ)> = \text{trace Re } DD^* + \text{trace Re } M(Z,qZ)$$

$$= \text{trace } DD^* + \text{trace Re } \{ZZ^* + (pq)\bar{Z}\bar{Z}^* (pq)^{-1}\}$$

$$= \text{trace } DD^* + 2 \text{ trace } ZZ^* > 0.$$

Now the pairing (14.2a) is nondegenerate, so

(14.2b) $n^* = \{f_{D,Z} : (D,Z) \in n\}$ where $f_{D,Z}(D',Z') = <(D,Z), (D',Z')>$.

As before we will write $f_D = f_{D,0}$ and $g_Z = f_{0,Z}$.

The analog of Proposition 4.3 is

14.3. Proposition. *If* $D \in \mathfrak{so}^*(2s)$, *then we can fix a direct sum decomposition*

(14.4a) $$\mathbb{C}^{2s \times t} = \{Z \in \mathbb{C}^{2s \times t}: DZ = 0\} \oplus \mathbb{C}_D^{2s \times t}$$

subject to: if the matrix D *is semisimple then* $\mathbb{C}_D^{2s \times t} = D \cdot \mathbb{C}^{2s \times t}$.

If $f = f_{D,Z} \in \mathfrak{n}^*$ *then the isotropy algebra* $\mathfrak{n}_f = \{x \in \mathfrak{n}: f[x,\mathfrak{n}] = 0\}$ *is*

(14.4b) $$\mathfrak{n}_f = \mathfrak{so}^*(2s) + \{Z' \in \mathbb{C}^{2s \times t}: DZ' = 0\}$$

and the co-adjoint orbit is the affine subspace

(14.4c) $$\mathrm{Ad}^*(N) \cdot f = f + \{g_{DZ'}: Z' \in \mathbb{C}_D^{2s \times t}\}.$$

Proof.   Since $D \in \mathfrak{so}^*(2s)$ we have $Dpq = -(p \cdot {}^t D)q = pq \cdot {}^t D^* = pq\bar{D}$.
Given $Z' \in \mathbb{C}^{2s \times t}$, it follows that

trace Re $D \cdot \mathcal{L}(Z', Z'')^* = 0$ for all $Z'' \in \mathbb{C}^{2s \times t}$ if and only if $DZ' = 0$.

That gives (14.4b), and (14.4c) follows as in Proposition 4.3.

                                                                    q.e.d.

Since $\mathfrak{so}^*(2s)$ contains nonsingular matrices, such as p, there exists $f \in \mathfrak{n}^*$ with $\mathfrak{n}_f = \mathfrak{so}^*(2s)$, so

14.5. Corollary. $N_{2s;t}^*$ *has square integrable representations.*
If $f = f_{D,Z} \in \mathfrak{n}^*$ we make the now-standard definitions

(14.6a) $\delta_f = \{D' \in \mathfrak{so}^*(2s): f(D') = 0\}$,   central ideal in $\mathfrak{n}$;

(14.6b) $\mathfrak{m}_f = \mathfrak{n}_f/\delta_f$ quotient algebra and $p: \mathfrak{n} \to \mathfrak{m}_f$ projection;

(14.6c) $\mathfrak{h}_f = p(\mathfrak{so}^*(2s) + C_D^{2s \times t})$ with $C_D^{2s \times t}$ as in (14.4a);   and

(14.6d) $\alpha_f = p(\delta_f + \{Z' \in C^{2s \times t}: DZ' = 0\})$.

As in §4, $\mathfrak{m}_f = \mathfrak{h}_f \oplus \alpha_f$, direct sum of ideals. If we set

(14.7a) M_f: the simply connected nilpotent Lie group for $\mathfrak{m}_f$, and

(14.7b) H_f: and A_f: the analytic subgroups for $\mathfrak{h}_f$ and α_f,

then we have

(14.7c) $M_f = H_f \times A_f$.

A_f is a complex vector group and the possibilities for H_f are

(14.8a) H_f is a Heisenberg group of real dimension $\geqslant 3$ ($D \neq 0$, $t > 0$),

(14.8b) H_f is a 1-dimensional real vector group ($D \neq 0$, $t = 0$),

(14.8c) H_f is the trivial group ($D = 0$).

The unitary representation class $[\pi_f] \in N$ associated to $Ad^*(N)$ f by the Kirillov Theory is

(14.9a) $[\pi_f] = [\bar{\pi}_f \cdot p]$ where $[\bar{\pi}_f] \in \hat{M}_f$ is the class for $\bar{f}$ with
             $f = \bar{f} \cdot p$,

and (14.7c) tells us that

(14.9b) $[\bar{\pi}_f] = [\eta_f \otimes \alpha_f]$ where $[\eta_f] \in \hat{H}_f$ and $\alpha_f \in \hat{A}_f$.

Evidently $\alpha_f = e^{ig_Z}: \exp(\delta_f + Z') \to e^{ig_Z(Z')} = e^{i \text{ trace Re } \mathcal{M}(Z,Z')}$ and

η_f acts on $p(\mathfrak{so}*(2s))$ by the unitary character

$e^{if_D}: \exp(\delta_f + D') \to e^{if_D(D')} = e^{i \text{ trace Re } DD'*}$. According to the three

cases of (14.8),

(14.10a) H_f is Heisenberg and $\eta_f \in \hat{H}_f$ has central character $e^{if_D} \neq 1$,

(14.10b) $H_f \cong R^1$ and η_f is its unitary character $e^{if_D} \neq 1$, or

(14.10c) H_f and η_f are trivial.

We summarize, describing $\hat{N}$:

 <u>14.11. Theorem.</u> *Let* $N = N_{2s;t}^*$. *Then* $\hat{N}$ *consists of the classes*

$[\pi_{D,Z}] = [\pi_{f_{D,Z}}]$ *given as follows.*

(14.12a) $D \in \mathfrak{so}*(2s)$ *and* $Z \in C^{2s \times t}$ *modulo* $D \cdot C^{2s \times t}$,

(14.12b) $p: N \to M_f = H_f \times A_f$ *as in* (14.6) *and* (14.7) *with* $f = f_{D,Z}$,

(14.12c) α_f *is the unitary character* e^{ig_Z} *on* A_f,

(14.12d) $[\eta_f] \in \hat{H}_f$ *is the class given by* (14.8) *and* (14.10), *and*

(14.12e) $[\pi_{D,Z}] = [(\eta_f \otimes \alpha_f) \cdot p] \in \hat{N}$.

Furthermore,

   (i) $[\pi_{D,Z}] = [\pi_{D',Z'}]$ *if and only if* $D = D'$ *and* $Z - Z' \in D \cdot C^{2s \times t}$,

   (ii) *the central character of* $[\pi_{D,Z}]$ *restricts to* e^{if_D} *on* $\mathfrak{so}*(2s)$,

   (iii) *if* $D \neq 0$ *and* $t > 0$ *then* $[\pi_{D,Z}]$ *is infinite dimensional,*

   (iv) *if* $D = 0$ *or* $t = 0$ *then* $[\pi_{D,Z}]$ *is a unitary character.*

Finally, the Plancheral measure on $\hat{N}$ *is concentrated on* $\{[\pi_{D,Z}]:$ *the matrix* $D \in \mathfrak{so}*(2s)$ *is invertible}.*

We calculate the action of $SO*(2t) \times GL(s;Q)$ on $\mathfrak{n}*$ in terms of (14.2):

14.13 Lemma. $SO*(2t) \times GL(s;Q)$ *acts on* $\mathfrak{n}*$ *by*

$$(g,\gamma) \cdot f_{D,Z} = f_{\gamma'D\gamma^{-1}, \gamma'Zg} .$$

Proof. Let $Z'' = Z'_{(g^{-1})}$ we calculate $[(g,\gamma)f_{D,Z}](D',Z') =$

$f_{D,Z}[(g,\gamma)^{-1}(D',Z')] = f_{D,Z}(\gamma^{-1}D'\gamma, \gamma^{-1}Z'') = f_D(\gamma^{-1}D'\gamma) + g_Z(\gamma^{-1}Z'')$. Now

$$f_D(\gamma^{-1}D'\gamma) = \text{trace } Re(D\gamma^{-1}D'\gamma) = \text{trace } Re(\gamma'D\gamma^{-1}D') = f_{\gamma'D\gamma^{-1}}(D')$$

and, using (13.7b) and the commutation rules on p, q and γ,

$$g_Z(\gamma^{-1}Z'') = \text{trace } Re\ \mathcal{M}(Z, \gamma^{-1}Z'')$$

$$= \text{trace } Re\ \{(Z, pq\bar{Z}) \cdot \tilde{q} \cdot \{(\gamma^{-1} \cdot Z', pq\bar{\gamma}^{-1} \cdot \bar{Z}')g\}^* \cdot q\}$$

$$= \text{trace } Re\ \{(Z, pq\bar{Z}) \cdot q \cdot g* \cdot (Z', pq\bar{Z}') \cdot {}^{t}\bar{\gamma}^{-1} \cdot q\}$$

$$= \text{trace } Re\ \{(Z, pq\bar{Z}) \cdot g^{-1} \cdot \tilde{q} \cdot (Z', pq\bar{Z}') \cdot q \cdot \gamma'\}$$

$$= \text{trace } Re\ \{(\gamma'Z, pq\bar{\gamma}'\bar{Z}) \cdot g^{-1} \cdot \tilde{q} \cdot (Z', pq\bar{Z}') \cdot q\} = g_{\gamma'Z_g}(Z').$$

Thus $[(g,\gamma')f_{D,Z}](D',Z') = f_{\gamma'D\gamma^{-1}}(D') + g_{\gamma'Z_g}(Z') = f_{\gamma'D\gamma^{-1}, \gamma'Z_g}(D',Z')$

identically in (D',Z').   *q.e.d.*

It is convenient to collect some mildly exotic linear algebra at this point.

   14.14. Proposition. *There are precisely* $\frac{1}{2}(s+1)(s+2)$ *orbits for the co-adjoint action* $\gamma: f_D \to f_{\gamma' D \gamma^{-1}}$ *of* $GL(s;Q)$ *on the annihilator* $C^{2s \times t}$ *in* $\mathfrak{n}^*$. *Those orbits are represented by the* $f_{D_{k,\ell}}$, k *and* ℓ *non-negative integers such that* $0 \leqslant k + \ell \leqslant s$, *where*

$$D_{k,\ell} = \begin{bmatrix} A_{k,\ell} & 0 \\ 0 & -A_{k,\ell} \end{bmatrix} \quad \underline{with} \quad A_{k,\ell} = \begin{bmatrix} iI_k & 0 & 0 \\ 0 & -iI_\ell & 0 \\ 0 & 0 & 0 \end{bmatrix} \in C^{s \times s}.$$

Furthermore, the $GL(s;Q)$-*stabilizer of* $D_{k,\ell}$ *is isomorphic to* $Q^{(s-k-\ell) \times (k+\ell)} \cdot \{GL(s-k-\ell;Q) \times Sp(k,\ell)\}$.

   Proof. The identities $^t D \cdot p + pD = 0 = D^* \cdot q + qD$ defining $\mathfrak{so}^*(2s)$ are equivalent to

(14.15a) $(iDp) + {}^t(iDp) = 0$ and $(pq)(iDp)(pq)^{-1} = \overline{iDp}$

We put Q^s onto C^{2s} by $\sum e_\ell(a^\ell + b^\ell j) \to \sum e_\ell a^\ell + \sum e_{s+\ell} \overline{b}^\ell$ as usual, so that $A + Bj \in Q^{s \times s}$ corresponds to $\begin{bmatrix} A & B \\ -\overline{B} & \overline{A} \end{bmatrix} \in C^{2s \times 2s}$. Now (14.15a) says that $iDp = \begin{bmatrix} A & B \\ -\overline{B} & \overline{A} \end{bmatrix}$ with $A + {}^t A = 0$ and $B = B^*$.

   The $GL(s;Q)$-orbit structure of $\mathfrak{so}^*(2s)$ is not changed when we precede by the automorphism $\gamma \to \gamma' = p \cdot {}^t \gamma^{-1} \cdot p$ of $GL(s;Q)$. That done, $GL(s;Q)$ acts on $\mathfrak{so}^*(2s)$ by $\gamma: D \to \gamma D \gamma'^{-1} = \gamma Dp {}^t \gamma p$. This carries over to the action $\gamma: iDp \to \gamma \cdot iDp \cdot {}^t \gamma$ on antisymmetric matrices which come from $Q^{s \times s}$.

   Given $D \in \mathfrak{so}^*(2s)$, let d be the antisymmetric bilinear form on C^{2s} with matrix iDp in the standard basis. We are going to see that there is a change of basis from $GL(s;Q)$ such that, in the new basis,

(14.15b) d has matrix $\begin{bmatrix} 0 & B \\ -B & 0 \end{bmatrix}$ where $B = \begin{bmatrix} B' & 0 \\ 0 & 0 \end{bmatrix}$ and $B' = \begin{bmatrix} -I_k & 0 \\ 0 & I_\ell \end{bmatrix}$.

First, let $\{f_1,\ldots,f_r\} \subset C^{2s}$ be maximal for the property that the f_ℓ are linearly independent over Q with

$$d(f_\ell, f_\ell j) = 0 \text{ for } 1 \leq \ell \leq r \text{ and } d(f_k Q, f_\ell Q) = 0 \text{ for } k \neq \ell.$$

Then $\sum f_\ell Q$ is a d-nondegenerate subspace $C^{2r} \subset C^{2s}$, and we have a d-orthogonal splitting $C^{2s} = C^{2r} \oplus C^{2(s-r)}$ with $C^{2(s-r)}$ also quaternionic.

Second, if d is not identically zero on $C^{2(s-r)}$, we choose $h_1, h_2 \in C^{2(s-r)}$ with $d(h_1,h_2) = 1$. The h_i are independent/Q by maximality of our choice of $\{f_\ell\}$, so we have a subspace $U \cong C^4$ with basis $\{h_1,h_2;h_1 j, h_2 j\}$. Again by maximality of $\{f_\ell\}$, $d(h,hj) = 0$ on U, so

$$0 = d(h_1 + h_2, (h_1 + h_2)j) = d(h_1, h_2 j) + d(h_2, h_1 j)$$

$$= d(h_1, h_2 j) - d(h_1 j, h_2) = 2d(h_1, h_2 j).$$

Thus $d(h_1, h_2 j) = 0$, and similarly $d(h_2, h_1 j) = 0$, so now $d|_{U \times U}$ has matrix $\begin{bmatrix} J & 0 \\ 0 & J \end{bmatrix}$, $J = \begin{bmatrix} 0 & 1 \\ -1 & 0 \end{bmatrix}$, in our basis of U. We then have $d(h_1 + h_2 i, (h_1 + h_2 i)j) = d(h_1 + h_2 i, h_1 j - (h_2 j)i) = 2$, contradicting maximality of $\{f_\ell\}$. Thus d is identically zero on $C^{2(s-r)}$.

We now have a quaternionic change of basis such that, in the new basis $\{f_1,\ldots,f_s;f_1 j,\ldots,f_s j\}$ of C^{2s}, d has matrix $\begin{bmatrix} 0 & B \\ -\bar{B} & 0 \end{bmatrix}$ where $B = \begin{bmatrix} B' & 0 \\ 0 & 0 \end{bmatrix}$ where $B' = \begin{bmatrix} b_1 & & \\ & \ddots & \\ & & b_r \end{bmatrix}$ and each $b_\ell \neq 0$. The matrix of d is antisymmetric, so $B = B^*$, that is $b_\ell = \pm\lambda_\ell^{-2}$ where λ_ℓ is nonzero and real. Replacing f_ℓ by $\lambda_\ell f_\ell$ we change b_ℓ to ± 1, and then permuting $\{f_1,\ldots,f_r\}$ we may suppose

$B' = \begin{bmatrix} -I_k & 0 \\ 0 & I_{r-k} \end{bmatrix}$ for some k. This proves (14.15b).

From (14.15b) we have an element $\gamma \in GL(s;Q)$ such that
$\gamma \cdot iDp \cdot {}^t\gamma = \begin{bmatrix} 0 & B \\ -B & 0 \end{bmatrix}$ with B as prescribed . Now

$\gamma \cdot D \cdot \gamma'^{-1} = -i(\gamma \cdot iDp \cdot {}^t\gamma)p = \begin{bmatrix} -iB & 0 \\ 0 & iB \end{bmatrix} = \begin{bmatrix} A & 0 \\ 0 & -A \end{bmatrix}$ where $A = \begin{bmatrix} A' & 0 \\ 0 & 0 \end{bmatrix}$ with

$A' = \begin{bmatrix} iI_k & 0 \\ 0 & -iI_\ell \end{bmatrix}$, $k + \ell = r \leqslant s$. Now, in particular, $D_{k,\ell} = \begin{bmatrix} A & 0 \\ 0 & -A \end{bmatrix}$ is in
the $GL(s;Q)$-orbit of D.

Let $H = \{\gamma \in GL(s;Q): \gamma \cdot D_{k,\ell} \cdot p \cdot {}^t\gamma \cdot p = D_{k',\ell'}\}$. We set

$B = B_{k,\ell} = \begin{bmatrix} -I_k & 0 & 0 \\ 0 & I_\ell & 0 \\ 0 & 0 & 0 \end{bmatrix}$ and $B' = B_{k',\ell'}$, and calculate

$$H = \left\{ \begin{bmatrix} E & F \\ -\bar{F} & \bar{E} \end{bmatrix} : \quad EBE^* + FBF^* = B' \quad \text{and} \quad EB \cdot {}^tF = FB \cdot {}^tE \right\}.$$

Another calculation shows that

$$(E + Fj)B(E + Fj)^* = B' \Leftrightarrow EBE^* + FBF^* = B' \quad \text{and} \quad EB \cdot {}^tF = FB \cdot {}^tE.$$

View $GL(s;Q)$ as quaternion matrices and denote hermitian forms on Q^s by

$$h(x,y) = -\sum_1^k x^{a-a}\bar{y} + \sum_{k+1}^{k+\ell} x^{b-b}\bar{y} \quad \text{and} \quad h'(x,y) = -\sum_1^{k'} x^{a-a}\bar{y} + \sum_{k'+1}^{k'+\ell'} x^{b-b}\bar{y} .$$

Now our calculation says that $H = \{\gamma \in GL(s;Q): \gamma^* \text{ sends h to h'} \}$.
Since rank and signature of hermitian forms on Q^s are $GL(s;Q)$-invariant,
we conclude that $D_{k,\ell}$ and $D_{k',\ell'}$ are in the same $GL(s;Q)$-orbit only
when $(k',\ell') = (k,\ell)$. This completes the determination of the orbit
structure in Proposition 14.14.

The stability subgroup of $GL(s;Q)$ at $D_{k,\ell}$ was just seen to be

(14.16a) $H_{k,\ell} = \left\{ \begin{bmatrix} E & F \\ -\bar{F} & \bar{E} \end{bmatrix} \in GL(2s;\mathbb{C}): EBE^* + FBF^* = B \text{ and } EB \cdot {}^tF = FB \cdot {}^tE \right\}$

where $B = B_{k,\ell}$. We further saw, from the quaternionic view,

(14.16b) $H_{k,\ell} \cong \{\gamma \in GL(s;\mathbb{Q}): \gamma^* \text{ preserves } h_{k,\ell} \text{ on } \mathbb{Q}^s\}.$

Now, as in Lemmas 6.1, 10.1 and 12.29,

$$H_{k,\ell} \cong \left\{ \begin{bmatrix} \sigma & \chi\tau \\ 0 & \tau \end{bmatrix} \in GL(s;\mathbb{Q}): \sigma \in Sp(k,\ell) = U(k,\ell;\mathbb{Q}) \right\}$$

$$\cong \{(\chi;\sigma,\tau) \in L_{s-k-\ell,k+\ell}(\mathbb{Q}): \sigma \in Sp(k,\ell)\}.$$

This completes the proof of Proposition 14.14.

$$q.e.d.$$

Now we go to the $SO^*(2t)$-stabilizer of $[\pi_{D,Z}] \in \hat{N}$, using Proposition 14.14 to modify the argument of Lemma 5.4.

14.17. Lemma. *Let* $D \in \mathfrak{so}^*(2s)$ *and* $Z \in \mathbb{C}^{2s \times t}$ *, and let* $L_{D,Z}$ *denote the* $SO^*(2t)$*-stabilizer of* $[\pi_{D,Z}]$ *. Then*

(14.18a) $L_{D,Z} = \{g \in SO^*(2t): Z_g - Z \in D \cdot \mathbb{C}^{2s \times t}\}.$

Choose $\gamma \in GL(s;\mathbb{Q})$ *with* $\gamma' \cdot D\gamma^{-1}$ *semisimple. Without changing* $[\pi_{D,Z}]$ *we add an element of* $D \cdot \mathbb{C}^{2s \times t}$ *to* Z *so that* $(\gamma' \cdot D\gamma^{-1})(\gamma'Z) = 0$ *. Assuming that normalization of* Z *, let* S_Z *denote the subspace of* $\mathbb{C}^{2t}$ *spanned by the columns of* ${}^t(Z, pq\bar{Z})$ *. Then*

(14.18b) $L_{D,Z} = \{g \in SO^*(2t): Z_g = Z\} =$
            $\{g \in SO^*(2t): gx = x \quad \text{for all } x \in S_Z\}$.

Proof. In view of Lemma 14.13, the description (14.4c) implies (14.18a).

Proposition 14.14 ensures the existence of $\gamma \in GL(s;Q)$ such that $\gamma' \cdot D\gamma^{-1}$ is semisimple, for the $D_{k,\ell}$ there are diagonal. Fixing one such γ, (14.4a) gives us

$$C^{2s \times t} = \{\gamma'Z' \in C^{2s \times t}: (\gamma'D\gamma^{-1})(\gamma'Z') = 0\} \oplus (\gamma'D\gamma^{-1}) \cdot \gamma' \cdot C^{2s \times t}.$$

As $GL(s;Q)$ acts by automorphisms, (14.4c) allows us to normalize Z as required. This action commutes with the action of $SO^*(2t)$, so $L_{D,Z} = L_{\gamma'D\gamma^{-1}, \gamma'Z}$. In view of our normalization the latter is $\{g \in SO^*(2t): \gamma'Z_g = \gamma'Z\} = \{g \in SO^*(2t): Z_g = Z\}$. As Z_g is defined by $(Z, pq\bar{Z}) \cdot g^{-1} = (Z_g, pq\overline{Z_g})$, and as $L_{D,Z}$ is a group, now $L_{D,Z} = \{g \in SO^*(2t): (Z, pq\bar{Z})g = (Z, pq\bar{Z})\} = \{g \in SO^*(2t): g(x) = x \text{ for all } x \in S_Z\}.$

q.e.d.

In the course of the proof of Lemma 6.1 we saw that Q^t has a nondegenerate skew-hermitian form, which we will call κ here, such that

(14.19a) $SO^*(2t) = \{g \in GL(t;Q): g \text{ preserves } \kappa\}.$

Under our standard $Q^t \to C^{2t}$, the matrix $-iI$ of κ goes over to $i\begin{bmatrix} -I & 0 \\ 0 & I \end{bmatrix}$ and the Q-subspaces of Q^t correspond to these subspaces $S \subset C^{2t}$ such that, for some $s \geqslant 0$,

(14.19b) S is spanned by the columns of ${}^t(Z, pq\bar{Z})$ for some $Z \in C^{2s \times t}$.

Given a subspace $S \subset C^{2t}$ as in (14.19b), let S also denote the corresponding subspace of Q^t, and let $S^{\perp}$ denote its orthocomplement relative to the form κ of (14.19a).

14.20. Lemma. *Let* $Z \in C^{2s \times t}$ *and let* $S = S_Z$, *the subspace of* C^{2t}
spanned by the columns of $^t(Z, pq\bar{Z})$. *Define integers* $c, u \geqslant 0$ *by*
$c = \dim_Q(S \cap S^\perp)$ *and* $S \cong (S \cap S^\perp) \oplus Q^u$. *Then* $\{g \in SO^*(2t): gx = x$
for all $x \in S\} \cong G^*_{2c; t-u-2c}$.

Proof. $S = T \oplus U$, sum of quaternionic subspaces, where $T = S \cap S^\perp$
and κ is nondegenerate on U. So $c = \dim T$, $u = \dim U$, $U^\perp = Q^{t-u}$,
and $g \to g|_{U^\perp}$ induces

$$\{g \in SO^*(2t): g|_S = \text{identity}\} \cong \{g' \in SO^*(2t-2u): g'|_T = \text{identity}\} .$$

The latter group is isomorphic to $G^*_{2c; t-u-2c}$ by Theorem 13.9.

$q.e.d.$

Now we extend $[\pi_{D,Z}]$ from N to $N \cdot L_{D,Z}$.

14.21. Proposition. *Let* $[\pi_{D,Z}] \in (N^*_{2s;t})^\wedge$ *as in Theorem* **14.11**.
Then $\pi_{D,Z}$ *extends to a unitary representation* $\tilde{\pi}_{D,Z}$ *of* $N^*_{2s;t} \cdot L_{D,Z}$
on the same Hilbert space.

Proof. Using Proposition 14.14 we replace (D, Z) by $(\gamma' D \gamma^{-1}, \gamma' Z)$
for appropriate $\gamma \in GL(s; Q)$, and then may suppose $D = \begin{bmatrix} A & 0 \\ 0 & -A \end{bmatrix}$ with
$A = \begin{bmatrix} A' & 0 \\ 0 & 0 \end{bmatrix}$ and $A' = \begin{bmatrix} iI_k & 0 \\ 0 & -iI_\ell \end{bmatrix}$. That done, we further replace Z within
$Z = D \cdot C^{2s \times t}$ so that $DZ = 0$.

Write the elements of $C^{2s \times t}$ in the form $\begin{bmatrix} Z' \\ Z'' \end{bmatrix}$ where $Z', Z'' \in C^{s \times t}$.
Using (13.7b) we see that our real $SO^*(2t)$-invariant inner product on
$C^{2s \times t}$ is given by

$$\langle Z_1, Z_2 \rangle = \text{trace Re } \mathcal{M}(Z_1, Z_2) = \text{trace Re } (Z_1, pq\overline{Z_1})\tilde{q}(Z_2, pq\overline{Z_2})^* q$$

$$= \text{trace Re } \begin{bmatrix} Z_1' & \overline{Z}_1'' \\ Z_1'' & -\overline{Z}_1' \end{bmatrix} \cdot \begin{bmatrix} -I & 0 \\ 0 & I \end{bmatrix} \cdot \begin{bmatrix} Z_2'^* & Z_2''^* \\ \overline{Z}_2''^* & -\overline{Z}_2'^* \end{bmatrix} \cdot \begin{bmatrix} -I & 0 \\ 0 & I \end{bmatrix}$$

$$= \text{trace Re } \begin{bmatrix} Z_1' Z_2'^* - \overline{Z}_1'' \overline{Z}_2''^* & \overline{} \\ \underline{} & -Z_1'' Z_2''^* + \overline{Z}_1' \overline{Z}_2'^* \end{bmatrix}$$

$$= 2 \text{ trace Re } \{Z_1' Z_2'^* - Z_1'' Z_2''^*\}.$$

Write x^u for the uth row of Z' and y^v for the vth row of Z''. Then $\begin{bmatrix} Z' \\ Z'' \end{bmatrix} \in D \cdot \mathbb{C}^{2s \times t}$ if and only if $x^w = 0 = y^w$ for $k + \ell < w \leq s$. Now for appropriate real $a_0 = a_0(D)$,

$$\beta(aD, \begin{bmatrix} Z' \\ Z'' \end{bmatrix}) = (iaa_0; x^1, \ldots, x^k, y^{k+1}, \ldots, y^{k+\ell}, y^1, \ldots, y^k, x^{k+1}, \ldots, x^{k+\ell})$$

induces an isomorphism from H_f, $f = f_{D,Z}$, onto the group $N_{1;(k+\ell)t,(k+\ell)t}(\mathbb{C})$ defined in (2.5). Equivariant to this, in view of the form of $\langle Z_1, Z_2 \rangle$, we have an injective homomorphism $\tilde{\beta} : SO^*(2t) \to U((k+\ell)t, (k+\ell)t)$. Now Proposition 5.8 says that η_f extends from H_f to $H_f \cdot L_{D,Z}$, and this implies the assertion of Proposition 14.21.

q.e.d.

Following Mackey's little-group method, the irreducible unitary representation classes for $G^*_{2s;t} = N^*_{2s;t} \cdot SO^*(2t)$ are just the unitarily induced classes

(14.22a) $[\pi_{D,Z,\gamma}] = [\text{Ind}_{N^*_{2s;t} \cdot L_{D,Z} \uparrow G^*_{2s;t}} (\tilde{\pi}_{D,Z} \otimes \gamma)]$

where

(14.22b) $[\gamma] \in \hat{L}_{D,Z}$ is extended to $N^*_{2s;t} \cdot L_{D,Z}$ by $\gamma(D', Z', g) = \gamma(g)$.

We have the usual equivalence condition for $[\pi_{D,Z,\gamma}]$ and $[\pi_{D',Z',\gamma'}]$. First, equivalence requires $D = D'$. That granted, suppose $D = D'$ semi-simplified, for example as in Proposition 14.14, by $\gamma \in GL(s;Q)$, and Z and Z' normalized by $DZ = 0 = D'Z'$. Then $[\pi_{D,Z,\gamma}] = [\pi_{D',Z',\gamma'}]$ if and only if there is an element $g \in SO^*(2t)$ such that

(14.22c) $D = D'$, $Z' = Z_g$ and $g' \to \gamma(gg'g^{-1})$ is equivalent to γ'.

We reformulate the equivalence condition (14.22c). Note that $Z' = Z_g$ for some $g \in SO^*(2t)$ if and only if $(Z',pq\bar{Z}') \in (Z,pq\bar{Z}) \cdot SO^*(2t)$. This happens just when some $g \in SO^*(2t)$ carries each column of ${}^t(Z,pq\bar{Z})$ to the corresponding column of ${}^t(Z',pq\bar{Z}')$. Let us set

S: subspace of C^{2t} spanned by the columns of ${}^t(Z,pq\bar{Z})$,

S': subspace of C^{2t} spanned by the columns of ${}^t(Z',pq\bar{Z}')$.

As in the discussion around (14.19), view S and S' as subspaces of Q^t. Let z_u and z'_u respectively denote the uth columns of ${}^t(Z,pq\bar{Z})$ and ${}^t(Z',pq\bar{Z}')$. The standard induction argument for Witt's Theorem is valid for (Q^t,κ). Thus there exists $g \in SO^*(2t)$ with $g(z_u) = z'_u$ if and only if $z_u \to z'_u$ defines a linear isomorphism of S onto S' and also $\kappa(z_u,z_v) = \kappa(z'_u,z'_v)$ for all u,v. The latter can be written $ZZ^* - (pq\bar{Z})(pq\bar{Z})^* = Z' \cdot Z'^* - (pq\bar{Z}')(pq\bar{Z}')^*$. In view of (13.7b) it is equivalent to $\mathcal{M}(Z,Z) = \mathcal{M}(Z',Z')$. Now

(14.23a) $Z' = Z_g$ with $g \in SO^*(2t)$ if and only if

(i) $\text{rank}(Z',pq\bar{Z}') = \text{rank}(Z,pq\bar{Z})$ and (ii) $\mathcal{M}(Z',Z') = \mathcal{M}(Z,Z)$.

Here note that the integers $c,u \geq 0$ in Lemma 14.20 are given by

(14.23b) $\text{rank}(Z, pq\bar{Z}) = 2(c+u)$ and $\text{rank } \mathcal{M}(Z,Z) = 2u.$

We formulate the preceding results and discussion as follows.

   14.24.   Theorem. $(G^{*}_{2s;t})^{\wedge}$ *is the disjoint union of non-empty subsets* $S(c,u;D)$, $D \in \mathfrak{so}^{*}(2s)$ *and* c,u *non-negative integers such that*

(*) $c + u + \frac{1}{2}\text{rank } D \leqslant s$ *and* $2c + u \leqslant t$

given as follows. If $D \in \mathfrak{so}^{*}(2s)$, *choose* $\beta \in GL(s;Q)$ *with* $\beta' \cdot D\beta^{-1}$ *semisimple. Let* $\mathcal{M}_{u,D}$ *denote the space of all* $2s \times 2s$ *complex matrices* M *of rank* $2u$ *such that*

   (i) M *is quaternionic:* $(pq)\bar{M}(pq)^{-1} = M$ *where* $pq = \begin{bmatrix} 0 & I \\ -I & 0 \end{bmatrix}$,

   (ii) M *is* q*-hermitian:* $Mq = qM^{*}$ *where* $q = \begin{bmatrix} 0 & I \\ I & 0 \end{bmatrix}$, *and*

   (iii) $(\beta' \cdot D\beta^{-1})(\beta' \cdot M\beta^{-1}) = 0.$

Then $S(c,u;D)$ *is parameterized by* $\mathcal{M}_{u,D} \times (G^{*}_{2c;t-u-2c})^{\wedge}$ *under*

$$(M,[\gamma]) \leftrightarrow [\pi_{D,Z,\gamma}] \text{ as in } (14.22)$$

where

$Z \in C^{2s \times t}$ *with* $\beta' D\beta^{-1} \cdot \beta' Z = 0$, $\mathcal{M}(Z,Z) = M$ *and* $\frac{1}{2}\text{rank}(Z, pq\bar{Z}) = c + u$

and where

$L_{D,Z}$ *is identified with its isomorph* $G^{*}_{2c;t-u-2c}$.

Proof. Let $[\pi_{D,Z,\gamma}] \in (G^*_{2s;t})^\wedge$ as in (14.22). Proposition 14.14

gives us $\beta \in GL(s;Q)$ with $\beta' \cdot D\beta^{-1}$ semisimple, and then Proposition 14.3

shows that we may assume $\beta' D\beta^{-1} \cdot \beta' Z = 0$. D is fixed, but Z is only

determined, now, up to a transformation $Z \to Z_g$ with $g \in SO^*(2t)$. In other

words, using (14.23), Z is free so long as $\text{rank}(Z,pq\bar{Z})$ and $M = \mathcal{M}(Z,Z)$

are not changed.

Let S be the column span of $^t(Z,pq\bar{Z})$ in C^{2t}, $c = \dim_Q(S \cap S^\perp)$

using the skew Q-hermitian form κ, and $u \geqslant 0$ such that $S \cong (S \cap S^\perp) \oplus Q^u$.

Then $\text{rank}(Z,pq\bar{Z}) = \dim_C S = 2(c+u)$. M is the complex matrix for $(\kappa(z_a,z_b))$

where the z_a are the columns of $^t(Z,pq\bar{Z})$ viewed as elements of Q^t, so

rank $M = \dim_C Q^u = 2u$. Use (13.7b) to express

$$M = \mathcal{M}(Z,Z) = \{ZZ^* - (pq)(\bar{Z}\bar{Z}^*)(pq)^{-1}\}q.$$

Then M is quaternionic because

$$(pq)\bar{M}(pq)^{-1} = (pq)\{\bar{Z}\bar{Z}^* - (pq)(ZZ^*)(pq)^{-1}\}(pq)^{-1} \cdot (pq)q(pq)^{-1} = M,$$

M is q-hermitian because

$$qM^* = (Mq)^* = \{ZZ^* - (pq)(\bar{Z}\bar{Z}^*)(pq)^{-1}\}^* = \{ZZ^* - (pq)(\bar{Z}\bar{Z}^*)(pq)^{-1}\} = Mq,$$

and we have the required relation with D,

$$\beta' D\beta^{-1} \cdot \beta' M\beta^{-1} = \{\beta' D\beta^{-1} \cdot \beta' Z - (pq)(\overline{\beta' D\beta^{-1} \cdot \beta' Z \cdot Z}^*)(pq)^{-1}\}q\beta^{-1} = 0,$$

because $\beta' D\beta^{-1}$ is quaternionic and $\beta' D\beta^{-1} \cdot \beta' Z = 0$.

$\beta' D\beta^{-1} \cdot \beta' Z = 0$ gives $\beta' D\beta^{-1} \cdot (\beta' Z, pq\overline{\beta' Z}) = 0$, and so

$2(c+u) + \text{rank } D = \text{rank}(Z,pq\bar{Z}) + \text{rank } D = \text{rank}(\beta' Z, pq\overline{\beta' Z}) + \text{rank}(\beta' D\beta^{-1}) \leqslant 2s$.

And $S \cong (S \cap S^\perp) \oplus Q^u \subset Q^t$ gives $2c + u \leqslant t$.

Conversely let $D \in \mathfrak{so}^*(2s)$, let c and u be non-negative integers that satisfy (*), and let $M \in \mathcal{M}_{u,D}$. The normal form of Proposition 14.14 and the argument of the last paragraph of the proof of Theorem 9.33 give $Z \in C^{2s \times t}$ with $\beta'D\beta^{-1} \cdot \beta'Z = 0$, $\mathcal{M}(Z,Z) = M$ and $\frac{1}{2}\mathrm{rank}(Z,pq\bar{Z}) = c + u$; and so $(M,[\gamma])$ corresponds to a class $[\pi_{D,Z,\gamma}]$.

$$q.e.d.$$

In the manner described at the end of §5, Theorem 14.24 gives a complete description of $(G^*_{2s;t})^\wedge$ in terms of the $SO^*(2\ell)^\wedge$ for $0 < \ell \leqslant t$.

§15. Representations of the Maximal Parabolic Subgroups of $SO^*(2m)$.

We now combine the results of §§1 and 14 to write out the irreducible
unitary representations of the maximal parabolic subgroups of the real forms
$SO^*(2m)$ of complex orthogonal groups. These parabolic subgroups are the

$$P_E = \{g \in SO^*(2m): gE = E\} \cong P^*_{2s;m-2s} \quad \text{given by (13.8)}$$

where E is a doubly isotropic subspace of even dimension $2s > 0$ in C^{2m}.
If we view C^{2m} as quaternion space Q^m, then $SO^*(2m)$ is the $GL(m;Q)$-
stabilizer of a nondegenerate skew-hermitian form κ on Q^m, and E is
just a totally κ-isotropic quaternionic subspace with $\dim_Q E = s > 0$. Our
procedure for the unitary dual of $P_E = P^*_{2s;m-2s} = G^*_{2s;m-2s} \cdot GL(s;Q)$ is also
valid for its maximal unimodular subgroup $'P^*_{2s;m-2s} = G^*_{2s;m-2s} \cdot GL'(s;Q)$ and
for other classes of subgroups in which $GL(s;Q)$ is cut down in a way that
allows us to understand the corresponding little-groups.

For convenience and reference to §14 we set $t = m-2s$.

15.1. Lemma. *The* $GL(s;Q)$-*stabilizer of a class* $[\pi_{D,Z,\gamma}] \in (G^*_{2s;t})^{\hat{}}$
defined in (14.22) *is*

$$M_{D,Z} = \{\beta \in GL(s;Q): \beta'D\beta^{-1} = D \;\; \underline{and} \;\; \beta'Z - Z_g \in D \cdot C^{2s \times t} \;\; \underline{for\ some} \;\; g \in SO^*(2t)\}.$$

Recall Proposition 14.14 *and suppose that* f_D *is in the* $GL(s;Q)$-*orbit of*
$f_{D_{k,\ell}}$. *Assume* Z *normalized so that* $D_{k,\ell} \cdot \beta'Z = 0$ *where* $\beta'D\beta^{-1} = D_{k,\ell}$
and let $u = \frac{1}{2}\text{rank}\,\mathcal{M}(Z,Z)$. *Then*

(15.2a) $M_{D,Z} \cong Q^{(s-k-\ell) \times (k+\ell)} \cdot \{K \times Sp(k,\ell)\}$ *where*

(15.2b) $K = \{(z;e,f) \in L_{u,s-k-\ell-u}(Q): e \in SO^*(2u) \subset GL(u;Q)\}.$

{*In particular* $K \cong P^*_{2(s-k-\ell-u);u}/\text{SO}^*(2(s-k-\ell-u))$}.

Proof. The first assertion on $M_{D,Z}$ follows from Lemma 14.13, the equivalence condition (14.22c), and the fact that $GL(s;Q)$ centralizes $SO^*(2t)$ in $P^*_{2s;t}$.

Replace $[\pi_{D,Z,\gamma}]$ within its $GL(s;Q)$-orbit so that $D = D_{k,\ell}$ as in Proposition 14.14, and then replace Z within $Z + D \cdot C^{2s \times t}$ so that $DZ = 0$. Write $D = \begin{bmatrix} A & 0 \\ 0 & -A \end{bmatrix}$ where $A = A_{k,\ell} = \begin{bmatrix} A' & 0 \\ 0 & 0 \end{bmatrix}$ with $A' = \begin{bmatrix} iI_k & 0 \\ 0 & -iI_\ell \end{bmatrix}$. Express $\beta \in GL(s;Q)$ by $\beta' = \begin{bmatrix} E & F \\ -\bar{F} & \bar{E} \end{bmatrix}$, so $\beta^{-1} = p \cdot {}^t\beta' \cdot p = \begin{bmatrix} E^* & {}^tF \\ -\bar{F}^* & {}^t\bar{E} \end{bmatrix}$, and so

$$\beta' D \beta^{-1} = D \Leftrightarrow EAE^* + FAF^* = A \quad \text{and} \quad EA \cdot {}^tF = FA \cdot {}^tE.$$

That, of course, is the condition of the group $H_{k,\ell}$ of (14.16), given in quaternionic matrices as $\left\{ E + Fj = \begin{bmatrix} \sigma & \chi\tau \\ 0 & \tau \end{bmatrix} \in GL(s;Q): \sigma \in Sp(k,\ell) \right\}$. In other words,

$$H_{k,\ell} = \left\{ \begin{bmatrix} E & F \\ -\bar{F} & \bar{E} \end{bmatrix}': E = \begin{bmatrix} a & x \\ 0 & b \end{bmatrix} \text{ and } F = \begin{bmatrix} u & y \\ 0 & v \end{bmatrix} \text{ with } a + uj \in Sp(k,\ell) \right\}.$$

The form of D and the fact $DZ = 0$ tell us that $Z = \begin{bmatrix} 0 \\ Z'' \\ 0 \\ Z''' \end{bmatrix}$ and $D \cdot C^{2s \times t}$

consists of all $\begin{bmatrix} Z' \\ 0 \\ Z''' \\ 0 \end{bmatrix}$ where $Z', Z''' \in C^{(k+\ell) \times t}$ and $Z'', Z''' \in C^{(s-k-\ell) \times t}$.

Given $\beta \in H_{k,\ell}$ now, using (14.23),

$$\beta' Z - Z_g \in D \cdot C^{2s \times t} \quad \text{for some} \quad g \in SO^*(2t) \quad \text{if and only if}$$

$$\begin{bmatrix} b & v \\ -\bar{v} & \bar{b} \end{bmatrix} \quad \text{preserves the q-hermitian form} \quad \mathcal{M}\left(\begin{bmatrix} Z'' \\ Z''' \end{bmatrix}, \begin{bmatrix} Z'' \\ Z''' \end{bmatrix} \right).$$

That is the same as the condition that $b + vj \in GL(s-k-\ell;Q)$ preserve a skew-hermitian form of rank u on $Q^{s-k-\ell}$. In an appropriate basis, this condition is

$$b + vj = \begin{bmatrix} e & 0 \\ z & f \end{bmatrix} \text{ with } e \in SO^*(2u) \subset GL(u;Q).$$

We now have determined the isomorphism class of $M_{D,Z}$ as asserted in Lemma 15.1.

$$q.e.d.$$

An argument similar to that of Lemma 6.6 gives us

15.13. Lemma. *If* $[\pi_{D,Z,\gamma}] \in (G^*_{2s;t})^\wedge$, *then* $\pi_{D,Z,\gamma}$ *extends to a unitary representation* $\tilde{\pi}_{D,Z,\gamma}$ *of* $G^*_{2s;t} \cdot M_{D,Z}$ *on the same Hilbert space.*

According to the little-group method, $(P^*_{2s;t})^\wedge$ consists of the unitarily induced classes

$$(15.4a) \qquad [\pi_{D,Z,\gamma,\mu}] = [\text{Ind}_{G^*_{2s;t} \cdot M_{D,Z} \uparrow P^*_{2s;t}} (\tilde{\pi}_{D,Z,\gamma} \otimes \mu)]$$

where

$$(15.14b) \qquad D \in \mathfrak{so}^*(2s), \ Z \in C^{2s \times t}, \ [\gamma] \in \hat{L}_{D,Z}$$

and

$$(15.4c) \qquad [\mu] \in \hat{M}_{D,Z} \text{ is extended by } \mu(D',Z',g,\beta) = \mu(\beta).$$

Here classes $[\pi_{D,Z,\gamma,\mu}] = [\pi_{D',Z',\gamma',\mu'}]$ if and only if there exists $(g,\beta) \in SO^*(2t) \times GL(s;Q)$ such that

$$(15.5a) \qquad D' = \beta'D\beta^{-1} \text{ and } Z' - \beta'Z_g \in D' \cdot C^{2s \times t}$$

$$(15.5b) \qquad g' \rightarrow \gamma(gg'g^{-1}) \text{ is equivalent to } \gamma', \text{ and}$$

(15.5c) $\beta' \to \mu(\beta\beta'\beta^{-1})$ is equivalent to μ'.

With Lemma 15.1 in mind, we denote

(15.6a) $J^*_{k,\ell;u,v} = Q^{(k+\ell)\times(u+v)} \cdot \{Sp(k,\ell) \times L^*_{u,v}\}$

where

(15.6b) $L^*_{u,v} = \{(\chi;\alpha,\beta) \in L_{u,v}(Q): \beta \in SO^*(2v) \subset GL(v;Q)\}$.

Schematically, $J^*_{k,\ell;u,v}$ is given by
$$\begin{bmatrix} Sp(k,\ell) & \multicolumn{2}{c}{Q^{(k+\ell)\times(u+v)}} \\ & GL(u;Q) & Q^{u\times v} \\ 0 & 0 & SO^*(2v) \end{bmatrix}.$$

The principal result for maximal parabolic subgroups of $SO^*(2m)$ is

<u>15.7. Theorem</u>. $(P^*_{2s;t})^\wedge$ *is the* <u>*disjoint*</u> <u>*union*</u> <u>*of*</u> <u>*non-empty*</u> <u>*subsets*</u> $S(a,b;k,\ell)$ <u>*where*</u> a,b,k <u>*and*</u> ℓ <u>*are non-negative integers such that*</u>

(*) $a + b + k + \ell \leqslant s$ <u>*and*</u> $2a + b \leqslant t$

<u>*given*</u> <u>*as*</u> <u>*follows.*</u> <u>*Let*</u> $(a,b;k,\ell)$ <u>*satisfy*</u> (*) <u>*and choose*</u> $Z \in C^{2s\times t}$ <u>*such*</u> <u>*that*</u>

$$D_{k,\ell} \cdot Z = 0, \quad rank \ \mathcal{M}(Z,Z) = 2b \quad \underline{and} \quad rank(Z,pq\overline{Z}) = 2(a + b).$$

<u>*Then*</u> $S(a,b;k,\ell)$ <u>*is parameterized by*</u>

$$(G^*_{2a;t-b-2a})^\wedge \times (J^*_{k,\ell;s-k-\ell-b,b})^\wedge$$

<u>*under*</u>

$$([\gamma], [\mu]) \leftrightarrow [\pi_{D_{k,\ell},Z,\gamma,\mu}] \quad \underline{given\ by} \qquad (15.4)$$

where

$$L_{D_{k,\ell},Z} \quad \underline{is\ identified\ with\ its\ isomorph} \quad G^*_{2a;t-b-2a}$$

and

$$M_{D_{k,\ell},Z} \quad \underline{is\ identifed\ with\ its\ isomorph} \quad J^*_{k,\ell;s-k-\ell-b,b} \quad .$$

At this point, the parameterization is routinely verified. The $(G^*_{2a;t-b-2a})^\wedge$ are described in Theorem 14.24, so we turn to $(J^*_{k,\ell;s-k-\ell-b,b})^\wedge$.

The group $J^*_{k,\ell;u,v}$ will be viewed as the semidirect product

$$(15.8a) \qquad J^*_{k,\ell;u,v} = I^*_{k,\ell;u+v} \cdot L^*_{u,v}$$

where $L^*_{u,v} = \{(\chi;\alpha,\beta) \in L_{u,v}(Q): \beta \in SO^*(2v)\}$, as in (15.6b), and where

$$(15.8b) \qquad I^*_{k,\ell;u+v} = Q^{(k+\ell)\times(u+v)} \cdot Sp(k,\ell).$$

Looking back at (2.8) we see

$$(15.9a) \qquad I^*_{k,\ell;u+v} \cong G_{u+v;k,\ell}(Q)/Im\ Q^{(u+v)\times(u+v)}.$$

$Im\ Q^{(u+v)\times(u+v)}$ is central in the unipotent radical $N_{u+v;k,\ell}(Q)$ of $G_{u+v;k,\ell}(Q)$. In the notation (5.9), this tells us

$$(15.9b) \qquad (I^*_{k,\ell;u+v})^\wedge = \{[\pi_{D,Z,\gamma}] \in G_{u+v;k,\ell}(Q)^\wedge: D = 0\} .$$

Now Theorem 5.12 specializes to

    15.10. Proposition $(I^*_{k,\ell;u+v})^\wedge$ *is the disjoint union of non-empty sub-sets* $S_I(a,b,c)$, *where* a,b *and* c *are non-negative integers such that*

(*) $a + b + c \leqslant u + v, \quad a + c \leqslant k, \quad b + c \leqslant \ell$,

given as follows. Let $\mathcal{H}_{a,b}$ *denote the space of all* $(u+v) \times (u+v)$ *hermitian matrices* H *over* Q *such that* H *has "signature"* $(a,b,u+v-a-b)$. *Then* $S_I(a,b,c)$ *is parameterized by* $\mathcal{H}_{a,b} \times G_{c,k-c-a,\ell-c-b}(Q)^\wedge$ *under*

$$(H,[\gamma]) \leftrightarrow [\pi_{0,Z,\gamma}] \quad \text{in the notation } (5.9) \text{ and } (15.9)$$

where $Z \in Q^{(u+v) \times (k,\ell)}$ *with* $\mathcal{H}(Z,Z) = H$ *and* rank $Z = a + b + c$.

    The $L^*_{u,v}$-stabilizer of $[\pi_{0,Z,\gamma}] \in (I^*_{k,\ell;u+v})^\wedge$ is obtained by specializing Lemma 6.1 to the case $D = 0$. If $Z \in Q^{(u+v) \times (k,\ell)}$ with rank $Z = a+b+c$ and with $\mathcal{H}(Z,Z)$ of "signature" $(a,b,u+v-a-b)$, that stabilizer is

(15.11a) $M_Z = \{B \in L^*_{u,v} \subset GL(u+v;Q): \; B^*Z \in Z \cdot Sp(k,\ell)\}$.

Restricting the extension provided by Lemma 6.6, we see that

(15.11b)   every $[\pi_{0,Z,\gamma}] \in (I^*_{k,\ell;u+v})^\wedge$ extends to $I^*_{k,\ell;u+v} \cdot M_Z$.

Now the little-group method tells us

    15.12. Theorem. $(J^*_{k,\ell;u,v})^\wedge$ *is the disjoint union of non-empty sub-sets* $S_J(c;\{H\})$ *where* a, b *and* c *are non-negative integers such that*

$$a + b + c \leqslant u + v, \quad a + c \leqslant k, \quad b + c \leqslant \ell$$

and where $\{H\}$ _is an_ $L^*_{u,v}$-_equivalence class of_ $(u+v) \times (u+v)$ _hermitian matrices_ H _over_ Q _of "signature"_ $(a,b,u+v-a-b)$. $S_J(c;\{H\})$ _is parameterized by_

$$G_{c;k-c-a,\ell-c-b}(Q)^{\wedge} \times \hat{M}_Z$$

under

$$([\gamma],[\mu]) \leftrightarrow [\mathrm{Ind}_{I^*_{k,\ell;u+v}\cdot M_Z \uparrow J^*_{k,\ell;u,v}} (\tilde{\pi}_{0,Z,\gamma} \otimes \mu)]$$

where $Z \in Q^{(u+v) \times (k,\ell)}$ _with_ $\mathcal{H}(Z,Z) = H$ _and_ $\mathrm{rank}\ Z = a + b + c$.

This reduces the determination of $(J^*_{k,\ell;u,v})^{\wedge}$ to the determination of the $\hat{M}_Z$. As before, we will see that the M_Z are groups of similar type, but with smaller matrices, so, by recursion on the degree of the matrices, we will be able to assume the $\hat{M}_Z$ known.

To complete the discussion of $J^*_{k,\ell;u,v}$ we note that Q^{u+v} carries

(15.13a)   an hermitian form h of "signature" $(a,b,u+v-a-b)$

and

(15.13b)   a skew-hermitian form κ of rank v

such that

(15.13c) $M_Z = \{g \in GL(u+v;Q):\ g$ preserves both h and $\kappa\}$.

In effect, $L_{u,v}^* = \{g \in GL(u+v;Q): \ g \text{ preserves } \kappa\}$, and the condition
(15.11a) that $g^* Z \in Z \cdot Sp(k,\ell)$ simply says that g preserves the hermitian
form h with matrix $\mathcal{H}(Z,Z)$. Decompose $Q^{u+v} = U \oplus V \oplus W$ where

$\quad\quad U \oplus V$ is the null space of κ, so $\kappa|_{W \times W}$ is nondegenerate,

$\quad\quad U$ is the null space of $h|_{(U+V) \times (U+V)}$, so $h|_{V \times V}$ is nondegenerate,

$\quad\quad W$ is orthogonal to V relative to h.

Every $g \in M_Z$ preserves $U \oplus V$ and U, and sends W to $U \oplus W$. Now
$g \in M_Z$ is given schematically by

$$
\begin{bmatrix} g_{11} & g_{12} & g_{13} \\ 0 & g_{22} & 0 \\ 0 & 0 & g_{33} \end{bmatrix} \in \begin{bmatrix} GL(U) & U \otimes V^* & U \otimes W^* \\ 0 & Sp(h|_{V \times V}) & 0 \\ 0 & 0 & SO^*(\kappa|_{W \times W}) \end{bmatrix}
$$

where g_{13} and g_{33} are related by the condition that $h(gw,gw') = h(w,w')$
for all $w,w' \in W$. Set $n_1 = \dim U$, $n_2 = \dim V$, $n_3 = \dim W$, and
$(m_1,m_2) = $ signature $(h|_{V \times V})$. Then $M_Z \cong I'_{n_1,n_2;n_3} \cdot L_{n_1;m_1,m_2}(Q)$ where
$L_{n_1;m_1,m_2}(Q) = \{(\chi;\sigma,\tau) \in L_{n_1,n_2}(Q): \tau \in Sp(m_1,m_2)\}$ as in (6.9d), and
where $I'_{n_1,n_2;n_3}$ is the subgroup $(g|_{U+V} = \text{identity})$ of M_Z, contained
in the obvious way

$$
\begin{bmatrix} I & 0 & g_{13} \\ 0 & I & 0 \\ 0 & 0 & g_{33} \end{bmatrix} \rightarrow \begin{bmatrix} I & g_{13} \\ 0 & g_{33} \end{bmatrix} \rightarrow (g_{13}\, g_{33}^{-1},\, g_{33})
$$

in $Q^{n_1 \times n_3} \cdot SO^*(2n_3)$. This gives us enough information to determine $\hat{M}_Z$
given specific h and κ.

This completes our determination of the unitary duals of the maximal parabolic subgroups of the classical real simple Lie groups $SO^*(2m)$.

Appendix: Induced Representations

In this appendix, we collect the basic definitions and theorems concerning induced representations and the little-group method, in the somewhat restricted generality required in this Memoir. The reader is referred to Mackey ([7], [8], [9], [10], [11]; see [12]), Moore ([1], [13], [14]) and Duflo [2, Ch. V] for complete proofs and for the most general results. See Lipsman [6] for a more extensive sketch of group representation theory.

A. I. Definition of Induced Representation.

G is a separable locally compact group, $dg = d\mu_G(g)$ denotes its left Haar measure, and Δ_G is the modular function. So

$$\int_G f(xg)\,dg = \int_G f(g)\,dg = \int_G f(g^{-1})\Delta_G(g^{-1})\,dg$$

and

$$\int_G f(xgx^{-1})\,dg = \Delta_G(x)\int_G f(g)\,dg = \int_G f(gx^{-1})\,dg$$

for $x \in G$ and $f \in C_c(G)$, the space of continuous compactly supported functions $G \to \mathbb{C}$.

Let L be a closed subgroup, $d\ell = d\mu_L(\ell)$ its left Haar measure, and Δ_L its modular function. Whenever η is a weakly continuous homomorphism from L to the bounded linear operators on a Hilbert space H_η , we denote

$$(A.1) \qquad C_c(G/L,\eta) = \left\{ f\colon G \to H_\eta : \begin{array}{l} \text{(i)} \quad f \text{ is continuous} \\ \text{(ii)} \quad f \text{ is compactly supported mod L} \\ \text{(iii)} \quad f(g\ell) = \eta(\ell)^{-1}f(g) \text{ for } g\in G, \ \ell\in L \end{array} \right\}$$

For the moment we only look at $C_c(G/L, \Delta_{G/L})$ where

(A.2) $\Delta_{G/L}: L \to R^+ = \{t \in R: t > 0\}$ by $\Delta_{G/L}(\ell) = \Delta_G(\ell)/\Delta_L(\ell)$.

The linear map

$$\tau: C_c(G) \to C_c(G/L, \Delta_{G/L}) \ \text{ by } \ (\tau f)(g) = \int_L f(g\ell)\Delta_{G/L}(\ell)d\ell$$

is surjective, and $\tau(f) = 0$ implies $\int_G f(g)dg = 0$. So integration on G induces a linear functional on $C_c(G/L, \Delta_{G/L})$, which we write as integration against a measure $d(gL) = d\mu_{G/L}(gL)$:

(A.3) $\int_{G/L} F(gL)d(gL) = \int_{G/L} F(x)d\mu_{G/L}(x)$ for $F \in C_c(G/L, \Delta_{G/L})$,

defined as $\int_G f(g)dg$ where $F = \tau(f)$. The functional (A.3) is our positive continuous G-invariant integral on G/L.

We digress to justify the notation (A.3) as an integral. Let $\pi: G \to G/L$ projection and choose $\varphi: G \to R$ continuous, values ≥ 0, such that

$$\text{each } \varphi_g: \ell \to \varphi(g\ell) \ \text{ is in } \ C_c(L) \ \text{ with } \ \int_L \varphi_g(\ell)d\ell > 0 \ .$$

Then $\pi(\varphi\mu_G)$ is a positive measure on G/L, $\int_{G/L}\Phi(x)d\pi(\varphi\mu_G)(x)$ $= \int_G \Phi(gL)\varphi(g)dg$ for $\Phi \in C_c(G/L)$. Set $\psi = \tau(\varphi)$, that is $\psi(g) = \int_L \varphi(g\ell)\Delta_{G/L}(\ell)d\ell$, so $\Phi \mapsto \Phi\psi$ is a bijection of $C_c(G/L)$ onto $C_c(G/L, \Delta_{G/L})$. Then $\tau(\Phi\varphi) = \Phi\psi$, so $\int_{G/L}\Phi(x)d\pi(\varphi\mu_G)(x) = \int_G \Phi(gL)\varphi(g)dg$ $= \int_{G/L}\tau(\Phi\varphi)(x)d\mu_{G/L}(x) = \int_{G/L}\Phi(x)\psi(x)d\mu_{G/L}(x)$.

Now let η be a (weakly continuous) unitary representation of L, and let H_η denote the representation space. Then $\eta \otimes \Delta_{G/L}^{1/2}$ still is a weakly continuous homomorphism $\ell \mapsto \Delta_{G/L}^{1/2}(\ell)\eta(\ell)$ from L to the bounded operators on H_η, so we have $C_c(G/L, \eta \otimes \Delta_{G/L}^{1/2})$ as in (A.2). If $F_1, F_2 \in C_c(G/L, \eta \otimes \Delta_{G/L}^{1/2})$ then, since η is unitary, the pointwise inner product satisfies

$$\langle F_1(g\ell), F_2(g\ell) \rangle_{H_\eta} = \Delta_{G/L}(\ell)^{-1} \langle F_1(g), F_2(g) \rangle_{H_\eta},$$

and so (A.3) gives us a global inner product

(A.4) $\langle F_1, F_2 \rangle = \int_{G/L} \langle F_1(x), F_2(x) \rangle_{H_\eta} \, d\mu_{G/L}(x).$

G has a natural unitary representation on the Hilbert space completion of $C_c(G/L, \eta \otimes \Delta_{G/L}^{1/2})$ with respect to (A.4), $[g(F)](g_0) = F(g^{-1}g_0)$, which is denoted and called

(A.5) $\mathrm{Ind}_{L\uparrow G}(\eta)$, representation of G induced by η.

It often is convenient to view $\mathrm{Ind}_{H\uparrow G}(\eta)$ as the action of G on L_2 sections of the Hilbert space bundle over G/L associated to $G \to G/L$ by the action $\eta \otimes \Delta_{G/L}^{1/2}$ of L on H_η. And of course everything is simplified when $\Delta_{G/L} = 1$.

An important example: if $\Delta_{G/L} = 1$, and if 1_L denotes the trivial 1-dimensional representation of L, then $\mathrm{Ind}_{L\uparrow G}(1_L)$ is the "left regular" representation of G on $L_2(G/L)$. In particular, $\mathrm{Ind}_{\{1\}\uparrow G}(1_{\{1\}})$ is the left regular representation of G.

The key technical result in working with induced representations is induction by stages: given closed subgroups $L \subset M \subset G$ one has

(A.6) $\mathrm{Ind}_{L\uparrow G}(\eta)$ is equivalent to $\mathrm{Ind}_{M\uparrow G}(\mathrm{Ind}_{L\uparrow M}(\eta)).$

A.II. Mackey's Little Group Method

If π is a unitary representation we write $[\pi]$ for its unitary equivalence class. If G is a locally compact group we write $\hat{G}$ for its unitary dual,

$$\hat{G} = \{[\pi]: \pi \text{ is an irreducible unitary representation of } G\}$$

with its usual Borel structure ([10]; or see [1] or [6]).

Let L be a closed normal subgroup of G. Then G acts on the representations of L by conjugation,

$$g: \eta \to \eta^g \quad \text{where} \quad \eta^g(\ell) = \eta(g^{-1}\ell g)$$

and this induces an action of G on $\hat{L}$. Given $[\eta] \in \hat{L}$, denote its stabilizer

$$(A.7) \qquad G_\eta = \{g \in G: \ell \to \eta(g^{-1}\ell g) \text{ is equivalent to } \eta\}$$

and consider the "extensions"

$$(A.8) \qquad \& (\eta) = \{[\psi] \in \hat{G}_\eta: \psi|_L \text{ is equivalent to a multiple of } \eta\} \ .$$

If G and L are type I, and if there is a Borel section to the action of G on $\hat{L}$, then the Mackey Little Group Theorem ([4], [5], [6], [8]; see [9]) says:

$$(A.9) \qquad \hat{G} = \{[\text{Ind}_{G_\eta \uparrow G}(\psi)]: [\eta] \in \hat{L} \text{ and } [\psi] \in \hat{G}_\eta\} \ ,$$

and

$$(A.10) \quad [\mathrm{Ind}_{G_\eta \uparrow G}(\psi)] = [\mathrm{Ind}_{G_{\eta'} \uparrow G}(\psi')] \Leftrightarrow \begin{cases} \text{(i)} \quad [\eta'] = [\eta^g] \text{ for some } g \in G \\ \text{(ii)} \quad \psi' \text{ is equivalent to} \\ \qquad x \to \psi(g^{-1}xg). \end{cases}$$

If it happens that (G,L) has the extension property

$$(A.11) \quad \text{if } [\eta] \in \hat{L} \text{ there exists } [\tilde{\eta}] \in \hat{G}_\eta \text{ with } \tilde{\eta}\big|_L = \eta,$$

then matters are simplified because

$$(A.12) \quad \&(\eta) = \{[\tilde{\eta} \otimes \mu]: [\mu] \in (G_\eta/L)^\wedge, \text{ lifted to } G_\eta\}.$$

Thus, under the conditions for (A.8) and (A.9),

$$(A.13) \quad \hat{G} = [\mathrm{Ind}_{G_\eta \uparrow G}(\tilde{\eta} \otimes \mu)]: [\eta] \in \hat{L} \text{ and } [\mu] \in (G_\eta/L)^\wedge\}.$$

Things simplify further when $G = L \cdot U$, semidirect product, with the extension property (A.11): then $G_\eta = L \cdot U_\eta$, semidirect with $U_\eta = U \cap G_\eta$, so

$$(A.12') \quad \&(\eta) = \{[\tilde{\eta} \otimes \mu]: [\mu] \in \hat{U}_\eta, \text{ lifted to } L \cdot U_\eta\}$$

and

$$(A.13') \quad \hat{G} = \{[\mathrm{Ind}_{L \cdot U_\eta \uparrow G}(\tilde{\eta} \otimes \mu)]: [\eta] \in \hat{L} \text{ and } [\mu] \in \hat{U}_\eta\}.$$

Except when dealing with real symplectic groups, $\hat{G}$ will always be given by (A.13') in this Memoir.

A.III. Cohomology and Projective Representations

When the extension property (A.11) is not available we are forced to use a chop-and-patch (cohomology) method to calculate the set $\mathscr{E}(\eta)$ of extensions of a class $[\eta] \in \hat{L}$.

Recall that a topological space is <u>polonais</u> if it is homeomorphic to a complete separable metric space. For example, separable locally compact groups, and the unitary groups of separable Hilbert spaces, are polonais. We will use

(A.14) $\begin{cases} \text{if } G \text{ is a polonais group and } K \text{ a closed subgroup then} \\ G \to G/K \text{ admits a Borel cross section.} \end{cases}$

in the unitary group case, and

(A.15) $\begin{cases} \text{if } G_i \text{ are separable metric groups, } G_1 \text{ polonais} \\ \text{then every Borel homomorphism } G_1 \to G_2 \text{ is continuous} \end{cases}$

in the locally compact case.

Let G and A be separable locally compact groups, A abelian but written multiplicatively. $G^{(q)}$ denotes $G \times \dots \times G$ (q times). We want the cases $q = 1$ and $q = 2$ of

(A.16) $H^q(G;A)$: cohomology based on Borel cochains $G^{(q)} \to A$.

This is a multiplicative abelian group. For example, using (A.15), one has

(A.17) $H^1(G;A) = Z^1(G;A)$: continuous homomorphisms $G \to A$.

Furthermore, the 2-cocycles form

(A.18) $Z^2(G;A)$: $\left\{ \begin{array}{l} \text{Borel maps} \quad \alpha\colon G \times G \to A \quad \text{with} \; \alpha(g,1) = \alpha(1,g) = 1, \\ \alpha(g,g')\alpha(gg',g'') = \alpha(g,g'g'')\alpha(g',g''). \end{array} \right\}$

The 2-coboundaries form the subgroup

(A.19) $B^2(G;A)$: $\left\{ \begin{array}{l} \text{Borel maps} \quad \alpha\colon G \times G \to A \quad \text{of the form} \; \alpha(g,g') \\ = \lambda(g)\lambda(g')\lambda(gg')^{-1} \quad \text{where} \; \lambda\colon G \to A \; \text{Borel}, \; \lambda(1) = 1 \end{array} \right\}.$

That gives the 2-cohomology

(A.20) $$H^2(G;A) = Z^2(G;A)/B^2(G;A).$$

One has the usual induced maps and exact sequences. For example, a continuous homomorphism $\varphi\colon A_1 \to A_2$ induces the "coefficient homomorphisms"

$$\varphi_*\colon H^q(G;A_1) \to H^q(G;A_2) \quad \text{by} \quad \varphi_*(\alpha) = \varphi \circ \alpha$$

and a continuous homomorphism $f\colon G_1 \to G_2$ induces

$$f^*\colon H^q(G_2;A) \to H^q(G_1;A) \quad \text{by} \quad f^*(\alpha) = \alpha \circ f$$

In applications more delicate than ours, one does all this for polonais G and separable metric A, and puts a polonais topology on the $H^q(G;A)$. See Moore ([13], [14], [15], [16]).

Let H be a Hilbert space. $U(H)$ denotes its unitary group, the circle group $\mathbb{C}' = \{\lambda \in \mathbb{C}\colon |\lambda| = 1\}$ is identified with the scalar operators in $U(H)$, and the quotient $PU(H) = U(H)/\mathbb{C}'$ is called the projective unitary group of H. Here we use the weak topology.

A continuous homomorphism $\bar{\pi}\colon G \to PU(H)$, H separable, is called a projective representation. The usual notions of irreducibility and unitary equivalence apply, and we write $[\bar{\pi}]$ for the unitary equivalence class of $\bar{\pi}$.

Let G be a separable locally compact group, H a separable Hilbert space, and $\bar{\pi}: G \to PU(H)$ a projective representation. By (A.14) we have a Borel section $\sigma: PU(H) \to U(H)$, $\sigma(1) = 1$, to the projection $p: U(H) \to PU(H)$. Denote

$$\pi = \sigma \circ \bar{\pi}: G \to U(H).$$

Glance back to (A.18) to see that it defines a 2-cocycle

$$\alpha: G \times G \to \mathbb{C}' \quad \text{by} \quad \alpha(g,g') = \pi(g)\pi(g')\pi(gg')^{-1}.$$

If $\sigma': PU(H) \to U(H)$ is another Borel section with $\sigma'(1) = 1$, then $\pi' = \sigma' \cdot \bar{\pi}: G \to U(H)$ is of the form $\pi'(g) = \lambda(g)\pi(g)$ where $\lambda: G \to \mathbb{C}'$, measurable, $\lambda(1) = 1$. Thus $\alpha: G \times G \to \mathbb{C}'$, defined by $\alpha'(g,g') = \pi'(g)\pi'(g')\pi'(gg')^{-1}$, satisfies

$$\alpha'(g,g') = \{\lambda(g)\lambda(g')\lambda(gg')^{-1}\}\alpha(g,g').$$

In other words, the cohomology class $[\alpha] \in H^2(G;\mathbb{C}')$ is independent of choice of σ. And evidently it only depends on the unitary equivalence class of $\bar{\pi}$. Denote

(A.21) $ProjRep(G) = \{[\bar{\pi}]: \bar{\pi}$ is a projective representation of $G\}$.

Now we have defined a map from representations to cohomology, which is Mackey's measurable variation on the classical Schur Multiplier construction in finite groups,

(A.22) $\mathscr{A}$: $\mathrm{Proj\,Rep}(G) \rightarrow H^2(G;\mathbb{C}')$

We say that a class $[\bar{\pi}] \in \mathrm{Proj\,Rep}(G)$ is <u>linear</u> if it contains a projective representation of the form $p\cdot\pi$ where π is an ordinary representation of G. Now $\mathscr{A}[\bar{\pi}]$ measures the failure of $[\bar{\pi}]$ to be linear. For if $\mathscr{A}[\bar{\pi}] = 1$ we have $\lambda: G \rightarrow \mathbb{C}'$ measurable, $\lambda(1) = 1$, such that

$$(\sigma\circ\bar{\pi})(g)\cdot(\sigma\circ\bar{\pi})(g') \circ (\sigma\circ\bar{\pi})(gg')^{-1} = \lambda(g)\lambda(g')\lambda(gg')^{-1}.$$

Then $\pi(g) = \lambda(g)^{-1}(\sigma\circ\bar{\pi})(g)$ is a Borel homomorphism $G \rightarrow U(H)$, continuous by (A15), such that $p\circ\pi \in [\bar{\pi}]$. And if $[\bar{\pi}]$ is linear, say $p\circ\pi \in [\bar{\pi}]$, then π is a Borel lift of $p\circ\pi$ and defines the trivial cocycle. In summary,

(A.23) $[\bar{\pi}] \in \mathrm{Proj\,Rep}(G)$ is linear if and only if $\mathscr{A}[\bar{\pi}] = 1$.

A. IV. Cocycle Representations and Extensions.

G is a separable locally compact group and $\alpha \in Z^2(G;\mathbb{C}')$, a 2-cocycle. By α-<u>*representation*</u> of G we mean a Borel map

$\pi: G \rightarrow U(H)$ such that $\pi(1) = 1$ and $\pi(g)\pi(g')\pi(gg')^{-1} = \alpha(g,g')$,

where H is a separable Hilbert space. Then π defines a projective representation $\bar{\pi} = p\circ\pi$. We say that an α-representation π' and an α'-representation π' are unitarily equivalent when the corresponding projective representations are equivalent. As noted above, that implies $[\alpha] = [\alpha']$. Denote

$\mathrm{Rep}_{[\alpha]}(G) = \{[\bar{\pi}] \in \mathrm{ProjRep}(G): \bar{\pi}$ defined by an α-representation$\}$.

Then   Proj Rep(G) is a disjoint union,

(A.24)   Proj Rep (G) $= \bigcup\limits_{[\alpha] \in H^2(G;\mathbb{C}')} \text{Rep}_{[\alpha]}(G)$.

If π_i is an α_i-representation then $\pi_1 \otimes \pi_2$, defined by
$(\pi_1 \otimes \pi_2)(g) = \pi_1(g) \otimes \pi_2(g)$, is an $\alpha_1 \alpha_2$ representation. So

(A.25)   if $[\pi_i] \in \text{Rep}_{[\alpha_i]}(G)$ then $[\pi_1 \otimes \pi_2] \in \text{Rep}_{[\alpha_1 \alpha_2]}(G)$.

We apply this machinery to the problem of extending representations.
Suppose that

(A.26) $\begin{cases} M \text{ is a separable locally compact group,} \\ L \text{ is a closed normal subgroup of } M , \\ [\eta] \in \hat{L} \text{ is stable under } M, \text{ i.e. } M_\eta = M. \end{cases}$

Given $m \in M$ now $\eta^m \colon \ell \mapsto \eta(m^{-1}\ell m)$ is equivalent to η. As η is
irreducible, this gives $\tilde{\eta}(m) \in U(H_\eta)$, unique modulo $\mathbb{C}'$, such that
$\eta(m^{-1}\ell m) = \tilde{\eta}(m)^{-1}\eta(\ell)\tilde{\eta}(m)$ for all $\ell \in L$. Setting $\bar{\eta}(m) = p\tilde{\eta}(m)$ we obtain

(A.27) $\bar{\eta} \colon M \to PU(H_\eta)$, projective representation.

Now $[\bar{\eta}] \in \text{Rep}_{[\alpha]}(M)$ where the cohomology class $[\alpha] = \mathcal{S}[\bar{\eta}]$ depends only
on $[\eta]$. Evidently, here α can be chosen so that $\alpha(L,L) = 1$. In fancier
language, the exact sequence $L \xrightarrow{i} M \xrightarrow{j} M/L$ gives an exact cohomology
sequence, part of which is

$$H^2(M/L;\mathbb{C}') \xrightarrow{j^*} H^2(M;\mathbb{C}') \xrightarrow{i^*} H^2(L;\mathbb{C}')$$

and $i^*\mathcal{S}[\bar{\eta}] = 1$. So $\mathcal{S}[\bar{\eta}]$ is in the image of j^*. Any class

(A.28) $m[\eta] \in H^2(M/L;\mathbb{C}')$ such that $j^* m[\eta] = \mathcal{S}[\bar{\eta}]$

is called the _Mackey obstruction_ to extension of $[\eta]$ from L to M. For,
in view of (A.23),

(A.29) $\begin{cases} \eta \text{ extends to a unitary representation } \tilde{\eta} \text{ of } M \text{ on the same} \\ \quad \text{Hilbert space } H_\eta \text{ if, and only if, } m[\eta] = 1. \end{cases}$

In general, if $[\mu] \in \text{Rep}_{[\beta]}(M/L)$ where $[\beta] = m[\eta]^{-1}$, and we lift μ to
M, then (A.23) and (A.25) show that $[\bar{\eta} \otimes \mu] \in \text{Proj Rep}(M)$ is linear.
The argument can be reversed if L is of type I. Thus, if L is of
type I, the set of extensions

$$\mathcal{E}(\eta) = \{[\psi] \in \hat{M}: \psi|_L \text{ equivalent to a multiple of } \eta\}$$

is given by

(A.30) $\mathcal{E}(\eta) = \{[\bar{\eta} \otimes \mu]: [\mu] \in \text{Rep}_{m[\eta]^{-1}}(M/L)\}.$

The solution (A.30) to extension problem gives a refinement of (A.9)
along the lines of the case (A.13) where (G,L) has the extension property:

(A.31) $\hat{G} = \begin{cases} & \text{(i)} \quad [\eta] \in \hat{L} \text{ and } \bar{\eta}: G_\eta \to \text{PU}(H_\eta) \text{ as in (A.27)} \\ [\text{Ind}_{G_\eta \uparrow G}(\psi): & \text{(ii)} \quad [\mu] \in \text{Rep}_{m[\eta]^{-1}}(G_\eta/L) \text{ and} \\ & \text{(iii)} \quad \psi \text{ is a unitary representation in } [\bar{\eta} \otimes \mu] \end{cases}$

References

[1] L. Auslander and C. C. Moore, "Unitary Representations of Solvable
     Lie Groups", Amer. Math. Soc. Memoir 62, 1966.

[2] P. Bernat, N. Conze, M. Duflo, M. Lévy-Nahas, M. Raïs, P. Renouard
     and M. Vergne, "Représentations des Groupes de Lie Résolubles",
     Dunod, Paris, 1972.

[3] A. Borel and J. Tits, Groupes réductifs, Publ. Sci. I.H.E.S. 27 (1965),
     55-150.

[4] A. A. Kirillov, Unitary representations of nilpotent Lie groups,
     Uspekhi Mat. Nauk. 17 (1962), 57-110 (in Russian). English trans-
     lation: Russian Math. Surveys 17 (1962), 53-104.

[5] R. P. Langlands, On the classification of irreducible representations of
     real algebraic groups. Preprint, Institute for Advanced Study,
     1973.

[6] R. Lipsman, "Group Representations", Springer Lecture Notes in Math.
     388, 1974.

[7] G. W. Mackey, Imprimitivity for representations of locally compact
     groups, I. Proc. Nat. Acad. Sci. U.S.A. 35 (1949), 537-545.

[8] _____, Induced representations of locally compact groups, I,
     Ann. of Math.(2)55 (1952), 101-139.

[9] _____, "Theory of Group Representations", University of Chicago
     lecture notes, 1955.

[10] _____, Borel structure in groups and their duals, Trans. Amer.
     Math. Soc. 85 (1957), 134-165.

[11] _____, Unitary representations of group extensions, I. Acta
     Math. 99 (1958), 265-311.

[12] _____, "Group Representations in Hilbert Space", Appendix to
     I. E. Segal, "Mathematical Problems in Relativistic Physics",
     Amer. Math. Soc., Providence, 1963.

[13] C. C. Moore, Extensions and low dimensional cohomology theory of
     locally compact groups, I. Trans. Amer. Math. Soc. 113 (1964),
     40-63.

[14] _____, Extensions and low dimensional cohomology theory of
     locally compact groups, II, Trans. Amer. Math. Soc. 113 (1964), 64-
     86.

[15] _____, Group extensions and cohomology for locally compact groups,
     III, Trans. Amer. Math. Soc., to appear.

[16] _____, Group extensions and cohomology for locally compact groups,
     IV, Trans. Amer. Math. Soc., to appear.

[17] C. C. Moore and J. A. Wolf, Square integrable representations of nil-
      potent groups, Trans. Amer. Math. Soc. 185 (1973), 445-462.

[18] L. Pukánszky,"Leçons sur les Représentations des Groupes", Dunod, Paris,
      1967.

[19] I. E. Segal, Chronogeometry and extreme distances, Astron. and Astro-
      physics 18 (1972), 143-148.

[20] _____, A variant of special relativity and extragalactic astron-
      omy, to appear.

[21] D. Shale, Linear symmetries of free boson fields, Trans. Amer. Math.
      Soc. 103 (1962), 149-167.

[22] S. Sternberg and J. A. Wolf, Charge conjugation and Segal's cosmology,
      Nuovo Cimento 28A (1975), 253-271.

[23] J. Tits, Espaces homogènes complexes compacts, Comment. Math. Helv. 37
      (1962-63), 111-120.

[24] A. Weil, Sur certains groupes d'opérateurs unitaires, Acta Math. 111
      (1964), 143-211.

[25] E. P. Wigner, On unitary representations of the inhomogeneous Lorentz
      group, Ann. of Math. 40 (1939), 149-204.

[26] J. A. Wolf, "Spaces of Constant Curvature", 3rd edition, Publish or
      Perish, Boston, 1973.

[27] _____, Fine structure of hermitian symmetric spaces, "Geometry and
      Analysis of Symmetric Spaces", Marcel Dekker, New York, 1972.

[28] _____, The action of a real semisimple Lie group on a complex flag
      manifold, I: Orbit structure and holomorphic arc components, Bull.
      Amer. Math. Soc. 75 (1969), 1121-1237.

[29] _____, The action of a real semisimple Lie group on a complex flag
      manifold, II: Unitary representations on partially holomorphic
      cohomology spaces. Amer. Math. Soc. Memoir 138, 1974.

[30] _____, Representations of certain semidirect product groups, J.
      Functional Analysis 19 (1975), 339-372.

[31] _____, Unitary representations of certain parabolic subgroups, in
      preparation.

The Hebrew University of Jerusalem
Tel Aviv University

Author's permanent address:

        Department of Mathematics
        University of California
        Berkeley, California 94720, U.S.A.